EXTEND
education

ReviseIB

Biology

TestPrep Workbook: DP Exam Practice

M000237955

A note from us

While every effort has been made to provide accurate advice on the assessments for this subject, the only authoritative and definitive source of guidance and information is published in the official subject guide, teacher support materials, specimen papers and associated content published by the IB. Please refer to these documents in the first instance for advice and guidance on your assessments.

Any exam-style questions in this book have been written to help you practise and revise your knowledge and understanding of the content before your exam. Remember that the actual exam questions may not look like this.

Sarah Bragg

HL
Higher Level

Published by Extend Education Ltd., Alma House, 73 Rodney Road, Cheltenham, UK GL50 1HT

www.extendeducation.com

The right of Sarah Bragg to be identified as the author of this work has been asserted by them with the Copyright, Designs and Patents Act 1988.

Reviewed by Nadia Jaradat

Typesetting by York Publishing Solutions Pvt. Ltd., INDIA

Cover photo by Yevhen Vitte on Shutterstock.com

First published 2022

26 25 24 23 22

10 9 8 7 6 5 4 3 2 1

ISBN 978-1-913121-02-0

Copyright notice

Other important information

A reminder that Extend Education is not in any way affiliated with the International Baccalaureate.

Many people have worked to create this book. We go through rigorous editorial processes, including separate answers checks and expert reviews of all content. However, we all make mistakes. So if you notice an error in the book, please let us know at info@extendeducation.co.uk so we can make sure it is corrected at the earliest possible opportunity.

If you are an educator with a passion for creating content and would like to write for us, please contact info@extendeducation.co.uk or write to us through the contact form on our website www.extendeducation.co.uk.

Permissions

Louisa Howard, p.11; Dartmouth Electron Microscope Facility, Dartmouth College, p. 12; Robert M. Hunt (CC BY 3.0), p.16; Shutterstock, Dimarion, p.17 & p.47; Wikimedia commons, Rosser1954, p.17 & p.48; Hussein A. Algahtani,a Abduljaleel P. Abdu,a Imad M. Khojah,b and Ali M. Al-Khathaamic (CC BY 2.5), p.27; Mrich (CC BY-SA 1.0), p.27; Deuterostome (CC BY-SA 3.0), p.28; Nephron (CC BY-SA 3.0), p.29; Vossman (CC BY-SA 3.0), p.30; Cadier en Keer (CC BY-SA 3.0) p.31; Ealbert17 (CC BY-SA 4.0), p.38; Jerry Crimson Mann (CC BY-SA 3.0), p.39; Adapted from Sadierath (CC BY-SA 4.0) p.39; www.scientificanimations.com (CC BY-SA 4.0), p.57; Ron Knight, Wikimedia commons (CC BY 2.0) p.58; Nandhu Kumar, p.60; Cybercobra at English Wikipedia (CC BY-SA 3.0), p.60; Toby Hudson (CC BY-SA 3.0), p.64; Acropora (CC BY 3.0), p.64; böhringer friedrich (CC BY-SA 2.5), p.64; Ian Alexander (CC BY-SA 4.0), p.64; © Hans Hillewaert (CC BY-SA 4.0), p.64; Matthias Tilly (CC BY 3.0), p.64; Jsarratt at English Wikipedia (CC BY 3.0), p.64; Charles J. Sharp, Sharp Photography (CC BY-SA 4.0), p.64; © Hans Hillewaert (CC BY-SA 4.0), p.64; Noah Elhardt (CC BY-SA 2.5), p.65; James Heilman, MD (CC BY-SA 3.0), p.68; adh30 revised work by DanielChangMD who revised original work of DestinyQx; Redrawn as SVG by xavax (CC BY-SA 4.0), p.69 & 151; albert kok (CC BY-SA 3.0) p.71; WillowW (CC BY-SA 3.0), p.73; Thomasione (CC BY-SA 3.0), p.74 & 95; CKRobinson (CC BY-SA 4.0), p.76; Yikrazuul (CC BY-SA 3.0), p.80; Nephron (CC BY-SA 3.0), p.81; iStock, blueringmedia, p.82; Robert M. Hunt (CC BY 3.0), p.87; Doc. RNDr. Josef Reischig, CSc. (CC BY-SA 3.0), p.94; SuperManu (CC BY-SA 2.5), p.95; brainmaps.org (CC BY 3.0), p.96; Gerry Carter (CC BY-SA 4.0), p.99; Brocken Inaglory (CC BY-SA 3.0), p.102; James Heilman, MD (CC BY-SA 3.0), p.109 & 163; Frevert U, Engelmann S, Zougbédé S, Stange J, Ng B, et al. Converted to SVG by Viacheslav Vtyurin who was hired to do so by User:Eug (CC BY 2.5), p.112; Zephyris (CC BY-SA 3.0), p.114; Cancer Research UK / Wikimedia Commons (CC BY-SA 4.0), p.118; Carl Davies, CSIRO (CC BY 3.0) p.119; OpenStax College (CC BY-SA 3.0), p.121; Ealbert17 (CC BY-SA 4.0), p.122; Nephron (CC BY-SA 3.0), p.123; Wapcaplet (CC BY-SA 3.0), p.127; Jon Houseman (CC BY-SA 4.0), p.128; Theresa knott (CC BY-SA 3.0), p.134; Carny (CC BY 2.5), p.135; Carny at Hebrew Wikipedia (CC BY 2.5), p.138; CNX OpenStax (CC BY 4.0), p.142; BirdPhotos.com (CC BY 3.0), p.145; Paul Whippey (CC BY-SA 3.0), p.145; Dani Kropivnik (CC BY-SA 3.0), p.146; Martin Talbot (CC BY 2.0) p.148; Adriana Zingone, Domenico D'Alelio, Maria Grazia Mazzocchi, Marina Montresor, Diana Sarno, LTER-MC team (CC BY-SA 4.0) p.148; Kulac (CC BY-SA 2.5) p.148; MichaelMaggs (CC BY-SA 3.0) p.148; Darekk2 (CC BY-SA 3.0), p.154.

CONTENTS

HOW TO USE THIS BOOK

This excellent exam practice book has been designed to help you prepare for your DP Biology HL exam. It is divided into three sections.

EXPLAIN

The EXPLAIN section gives you a rundown of your paper, including number of marks available, how much time you'll have and the assessment objectives (AOs) and command terms. There's also a handy checklist of your topics that you can use as a revision checklist.

SHOW

The SHOW section gives you some examples of different questions you will come across in the exam. It's designed to help you learn the question types and the kinds of answers you can give to get you the maximum number of marks.

TEST

This is your chance to try out what you've learned. The TEST section has full sets of exam-style practice papers filled with the same type and number of questions that you can expect to see in your exam. The first set of papers has a lot of helpful tips and suggestions for answering the questions. The middle set has more general advice - make sure you have revised before testing yourself with this set. The last set has no help at all. Not one single hint! Make sure you do this one a bit closer to your exam to check what else you might need to revise.

Set A
Paper 1, Paper 2 & Paper 3 (HL)

Presented with a lot of tips and guidance to help you to get to the correct answer and boost your confidence!

Use these papers early on in your revision.

Set B
Paper 1, Paper 2 & Paper 3 (HL)

Presented with fewer helpful suggestions so you have to rely on your revision before trying these.

Test yourself using these papers when you are a bit more confident.

Set C
Paper 1, Paper 2 & Paper 3 (HL)

Presented with space to add your own notes and no guidance – the perfect way to test whether you are exam ready.

Use these papers as close as you can to the exam.

All questions are presented with **ANSWERS** so you can check how you did in your practice papers.

Features

Take a look at some of the helpful features in these books that are designed to support you as you do your practice papers.

These will point you in the direction of the right answer!

These are general hints for answering the questions.

These are referred to as AOs all the way through this book

This box reminds you of the assessment objective being tested.

Beware of making common and easy-to-avoid mistakes!

These flag up common or easy-to-make mistakes that might cost you marks.

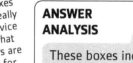

The command terms are like a clue to how you should answer your questions

COMMAND TERMS

These boxes outline what the command term is asking you to do.

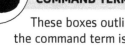

Link to TOK or Extended Essay!

These show you when the questions have other interdisciplinary links.

These boxes contain really useful advice about what examiners are looking for

ANSWER ANALYSIS

These boxes include advice on how to get the most possible marks for your answer.

KNOWING YOUR PAPER

Knowing the requirements of your exam papers is as important as your knowledge about the topic. The structure of the paper shouldn't be a surprise to you during the exam session. Being familiar with it will help you to answer the questions effectively.

How are you assessed?

You will sit three written papers for your Higher Level exams.

Paper 1	Paper 2	Paper 3
Multiple choice Answer all questions	Section A: Answer all questions Section B: Answer two questions	Section A: Answer all questions Section B: Answer all questions from one option
20% of final grade	36% of final grade	24% of final grade
40 marks	Section A: 40 marks Section B: 32 marks (Total 72 marks)	Section A: 15 marks Section B: 30 marks (Total 45 marks)
1 hour	2 hour 15 minutes	1 hour 15 minutes

> Try to keep your writing inside the boxes as these are what the examiner scans when they are marking. If you go outside the box, write 'continued on lined paper' at the bottom. Then, label the lined extra paper with the question number clearly.

Your assessment objectives

There are **three** assessment objectives for your IB biology HL exams. Make sure you are clear on what you are expected to demonstrate for each one.

Assessment objective	Command terms	Which questions test this?	Example question
Assessment objective 1: KNOWLEDGE AND COMPREHENSION OF SPECIFIED CONTENT	Define Draw Label List Measure State	Questions in the exam that test your understanding of AO1 are asking you to demonstrate knowledge and comprehension of different aspects of biology: • key terms and biological concepts • biological methodologies and techniques • how to communicate scientific information.	State the year ALOHA started measuring pCO2 and pH of seawater. **[1 mark]**
Assessment objective 2: APPLICATION AND ANALYSIS	Annotate Apply Calculate Describe Distinguish Estimate Identify Outline	Questions in the exam that test your understanding of AO2 are generally asking you to apply and analyse your knowledge. You may also be asked to use examples of biological methodologies and techniques, as well as how to communicate scientific knowledge, when answering a specific question.	Describe the trend in mortality when temperature is increased. **[2 marks]**

Assessment objective	Command terms	Which questions test this?	Example question
Assessment objective 3: SYNTHESIS AND EVALUATION	Analyse Comment Compare Construct Deduce Derive Design Determine Discuss Evaluate Explain Predict Show	Questions in the exam that test your understanding of AO3 are asking you to evaluate research questions and predictions. You will be asked to formulate and analyse primary and secondary data with explanations, using scientific methodologies and techniques.	Explain reasons for the change in the shell of *Limacina helicina* over the 45 days. **[3 marks]**

Before you begin your exam

Check that you have good knowledge and understanding of the topics before you take your exams. You could try sets A and B during your revision to see which areas you need to work on. Then do Set C after your revision to check that you are ready for the exam.

Higher Level topic checklist

Below is a checklist of core content for your DP biology HL exam. Put a tick in the box when you feel comfortable with the topic, when you have learned relevant methodologies to reference in your exam, and when you are confident in your understanding and use of terminologies.

Topic	Studied	Methodologies	Terminologies
Cell biology			
Introduction to cells			
Ultrastructure of cells			
Membrane structure			
Membrane transport			
The origin of cells			
Cell division			
Molecular biology			
Molecules to metabolism			
Water			
Carbohydrates and lipids			
Proteins			
Enzymes			
Structure of DNA and RNA			
DNA replication, transcription and translation			
Cell respiration			
Photosynthesis			
Genetics			
Genes			
Chromosomes			
Meiosis			
Inheritance			

Topic	Studied	Methodologies	Terminologies
Genetic modification and biotechnology			
Ecology			
Species, communities and ecosystems			
Energy flow			
Carbon cycling			
Climate change			
Evolution and biodiversity			
Evidence for evolution			
Natural selection			
Classification and biodiversity			
Cladistics			
Human physiology			
Digestion and absorption			
The blood system			
Defence against infectious disease			
Gas exchange			
Neurons and synapses			
Hormones, homeostasis and reproduction			
Nucleic acids			
DNA structure and replication			
Transcription and gene expression			
Translation			
Metabolism, cell respiration, and photosynthesis			
Metabolism			
Cell respiration			
Photosynthesis			
Plant biology			
Transport in the xylem of plants			
Transport in the phloem of plants			
Growth in plants			
Reproduction in plants			
Genetics and evolution			
Meiosis			
Inheritance			
Gene pools and speciation			
Animal physiology			
Antibody production and vaccination			
Movement			
The kidney and osmoregulation			
Sexual reproduction			

You will also cover one additional subject from the options below. You will only need to revise one.

Topic	Studied	Methodologies	Terminologies
Neurobiology and behaviour			
Neural development			
The human brain			
Perception of stimuli			
Innate and learned behaviour			
Neuropharmacology			
Ethology			
Biotechnology and bioinformatics			
Microbiology: organisms in industry			
Biotechnology in agriculture			
Environmental protection			
Medicine			
Bioinformatics			
Ecology and conservation			
Species and communities			
Communities and ecosystems			
Impacts of humans on ecosystems			
Conservation of biodiversity			
Population ecology			
Nitrogen and phosphorus cycles			
Human physiology			
Human nutrition			
Digestion			
Functions of the liver			
The heart			
Hormones and metabolism			
Transport of respiratory gases			

Here is a list of drawings you should know. Put a tick in the box when you feel comfortable you can draw, recognize and/or label the following.

Ultrastructure of a prokaryotic cell	
Ultrastructure of a eukaryotic cell (plant and animal)	
Draw the fluid mosaic model	
Draw water molecules to show hydrogen bonds as intermolecular forces.	
Draw diagrams showing the ring structure of alpha- and beta-D-glucose, D-ribose	
Draw diagrams of a saturated fat and generalized amino acid	
Draw diagrams to show the formation of a peptide bond	
Draw diagrams showing the structure of nucleotides of DNA and RNA using pentagons for the sugar, rectangles for the bases and circles for the phosphate	
Draw a short section of DNA with antiparallel strands, showing adenine pairing with thymine and cytosine pairing with guanine	
Draw the absorption spectrum for chlorophyll a and chlorophyll b	
Draw the action spectrum for photosynthesis	
Draw the stages of meiosis showing four gametes	

Construct Punnett grids for predicting the outcomes of monohybrid genetic crosses	
Construct dichotomous keys for use in identifying specimens	
Construct a diagram of the carbon cycle	
Construct a diagram of the digestive system	
Label the chambers, valves and blood vessels on a diagram of a heart	
Annotate the male and female reproductive system with structure and function	
Label a motor neuron	
Drawing the structure of primary xylem vessels in sections of stems based on microscope images	
Drawing internal structure of seeds	
Drawing of half-views of animal pollinated flowers	
Drawing diagrams to show chiasmata formed by crossing over	
Annotation of a diagram of the human elbow	
Drawing labelled diagrams of the structure of a sarcomere	
Drawing and labelling a diagram of the human kidney	
Annotation of diagrams of the nephron	

Make sure you have reviewed theories that have been covered on the syllabus.

Cell theory	
Miller-Urey experiment making amino acids from non-living molecules shows the first cells could have come from non-living matter	
Pasteur's theory of cells coming from pre-existing cells falsifies theory of spontaneous generation	
Wöhler's theory – artificially synthesizing urea in a lab falsifies theory of vitalism	
Davson–Danielli model replaced by Singer–Nicolson model	
Endosymbiotic theory	
Hydrogen bonds in water explain its polarity	
Watson and Crick developed the double helix model of DNA	
Meselson and Stahl's results supporting semi-conservative replication of DNA	
Evolution from natural selection – Darwin	
Molecular biology/proteins and DNA sequences used to reclassify plant families	
William Harvey's theories (on arteries and veins being connected with arteries flowing away from the heart and veins returning to it) falsifies Galen's idea (of arteries transporting heat and veins transporting blood)	
The chemiosmotic theory led to a paradigm shift in the field of bioenergetics	
The theory of punctuated equilibrium	

Are you familiar with the following practicals?

Calculate magnification, e.g. from scale bars	
Estimate osmolarity from graphs	
Investigate how a factor affects the rate of an enzyme-controlled reaction, e.g. temperature, pH or concentration of enzyme/substrate	
Identify R_f values from chromatographs	
Explain advantages and limitations of mesocosms as models for the environment	
Interpret spirometer traces including ventilation rate and tidal volume before and after exercise	
Measurement of transpiration rates using potometers	

What to do in your exam

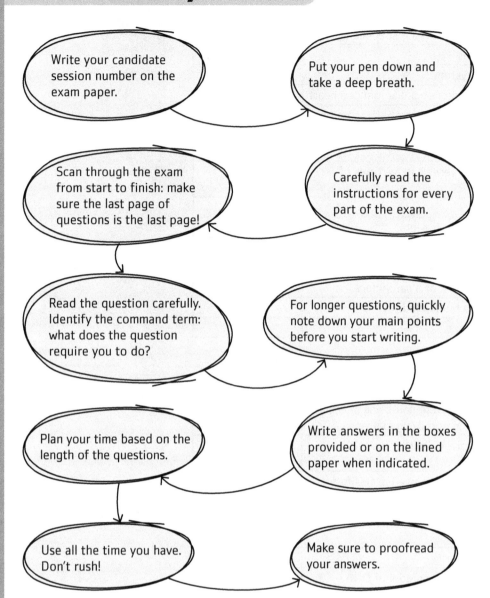

Write your candidate session number on the exam paper.

Put your pen down and take a deep breath.

Scan through the exam from start to finish: make sure the last page of questions is the last page!

Carefully read the instructions for every part of the exam.

Read the question carefully. Identify the command term: what does the question require you to do?

For longer questions, quickly note down your main points before you start writing.

Plan your time based on the length of the questions.

Write answers in the boxes provided or on the lined paper when indicated.

Use all the time you have. Don't rush!

Make sure to proofread your answers.

SHOWING WHAT YOU KNOW

This section of the book can help you to produce better answers to hit the top marks in your exam. Before looking at the answers, try answering the questions yourself first. Then compare your answers with the answers given.

Paper 1 examples

Welcome to Paper 1. This paper is made up of multiple-choice questions and covers the whole syllabus **excluding** the options. You have 1 hour to answer 40 questions (worth 40 marks). This means you should spend 1–2 minutes on each question. Leave the ones that you find more difficult until the end. Then spend the remaining time going back through the ones you found more difficult.

1. The cladogram shows the evolution of humans based on the gene for cytochrome c.

   Which species is most closely related to the olive baboon? **[1]**

   ☐ A. Rhesus monkey

   ☑ B. Gelada

   ☐ C. Golden snub-nosed monkey

   ☐ D. *Colobus angolensis palliatus*

   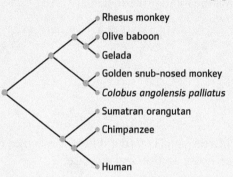

   - Rhesus monkey
   - Olive baboon
   - Gelada
   - Golden snub-nosed monkey
   - *Colobus angolensis palliatus*
   - Sumatran orangutan
   - Chimpanzee
   - Human

2. The image shows a transmission electron microscope image of a thin section cut through the pancreas (mammalian). Identify blood vessel X and cell component Y.

   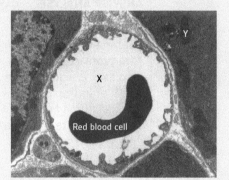

   What is the function of X and Y in the image? **[1]**

	X	Y
☐ A.	Artery	Rough endoplasmic reticulum
☐ B.	Artery	Mitochondria
☐ C.	Capillary	Rough endoplasmic reticulum
☑ D.	Capillary	Mitochondria

In your exam, the answers are on a separate answer sheet and must be filled out using pencil.

ANSWER ANALYSIS

The answer is B as they shared a common ancestor more recently. Next is the Rhesus monkey, then both the *colobus* and the golden snub-nosed monkey. Remember that branched names can rotate at the node.

If you have time left over, go back through and check your answers for any errors due to misinterpreting questions. Usually your first instinct is correct, so don't change your answer unless you are positive you have made a mistake. If in doubt, leave your original answer!

Some Paper 1 questions involve interpreting an image.

Some questions require you to choose the option that has the correct answer in both columns. In these questions, you can eliminate answers that are incorrect quickly and then get to the correct answer.

ANSWER ANALYSIS

The capillary lining consists of long, thin endothelial cells, connected by tight junctions. The red blood cell is large compared with the size of the lumen, so it is a capillary not an artery. A and B are eliminated. The answer is C or D. Rough endoplasmic reticulum is shown on the upper left side of the image. Y points to the mitochondrion as it is sausage or round in shape (depending on the plane of the cut) and it contains a highly folded inner membrane.

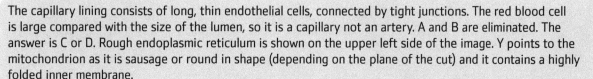

3. The image shows a transmission electron microscope of *Chlamydomonas reinhardtii*. This image of a thin section through a whole *Chlamydomonas* shows the nucleus, chloroplast, starch grains, vacuoles, mitochondria, eye spot and the cell wall.

Sometimes you need to calculate magnification from an image.

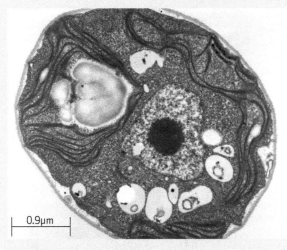

0.9μm

Calculate the approximate magnification of the image. **[1]**

☐ A. 18

☐ B. 180

☐ C. 1,800

☑ D. 18,000

ANSWER ANALYSIS

Remember: Magnification = image size/actual size.

Without a calculator you just need to approximate to make the maths easier. Image size = length of scale bar = 1.6 cm = 16 mm = 16,000 μm

Actual size from scale bar = 0.9 μm

0.9 is nearly 1. 16,000/1 = ×16,000 so choose the nearest answer (×18,000).

4. The graph shows the effect of increasing pH on four different enzymes in the human digestive system: pepsin, lipase, trypsin and amylase.

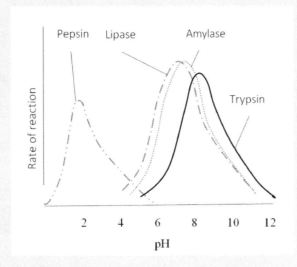

Sometimes you may need to interpret a graph.

ANSWER ANALYSIS

Both pepsin and trypsin are proteases, and the stomach is acidic so has a pH of 2. Trypsin would be denatured at pH2 but pepsin is at its optimum pH.

Which enzyme would be able to digest proteins in the stomach? **[1]**

☑ A. Pepsin

☐ B. Lipase

☐ C. Amylase

☐ D. Trypsin

Paper 2 (Section A) examples

Paper 2 is 2 hours and 15 minutes long and worth 72 marks. You will need to answer all the questions in Section A (worth 40 marks).

Paper 2, Section A, Question 1

1. A study in Hawaii has investigated how carbon dioxide emissions affect the ocean. Atmospheric CO_2 has been monitored since 1958 in Mauna Loa. Seawater carbon dioxide levels (measured as partial pressure pCO_2) and pH have been monitored at station ALOHA, which is north of Hawaii.

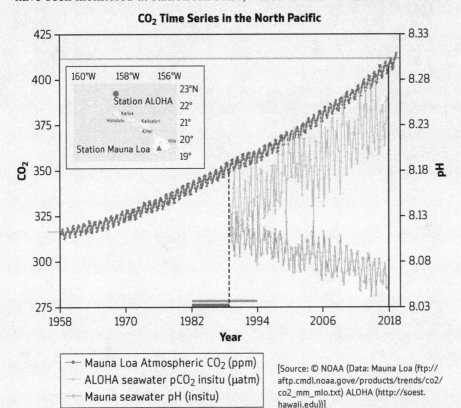

CO$_2$ Time Series in the North Pacific

- ● Mauna Loa Atmospheric CO_2 (ppm)
- ○ ALOHA seawater pCO_2 insitu (µatm)
- ● Mauna seawater pH (insitu)

[Source: © NOAA (Data: Mauna Loa (ftp://aftp.cmdl.noaa.gove/products/trends/co2/co2_mm_mlo.txt) ALOHA (http://soest.hawaii.edu))]

(a) State the year ALOHA started measuring pCO_2 and pH of seawater. **[1]**

14/24 × 12 = 7 years into the interval.

1982 + 7 = 1989

(b) Calculate the percentage increase in atmospheric CO_2 concentration between 1958 and 2018. **[1]**

(412 − 316) / 316 × 100 = 30%

Question 1 is an unseen data interpretation question with many sub questions that are connected together with a similar theme. Aim to spend about 30 minutes on question 1. Each mark should take about 2 minutes.

ANSWER ANALYSIS

To help identify the theme look in the text at the start of the question. This will give you clues for any later 'explain' or 'suggest' questions near the end of question 1. The question will check you can apply your scientific knowledge and analysis skills to an unseen data-based question. The questions are usually taken from scientific journals that link to the syllabus. Let's look at some examples.

Greenhouse gases such as methane and carbon dioxide absorb the infrared radiation (heat) and re-radiate the heat. The more greenhouse gases present, the more heat is re-radiated and the hotter the atmosphere becomes.

Draw a vertical line down from where they first appear. Then measure the distance of the line from 1982 = 14 mm in an interval that is 24 mm long and covers 12 years.

Check you are looking at the correct line and the CO_2 axis. Accurately read off using the same technique as above.

Percentage change = (change/original) × 100

(c) Discuss the evidence that the atmospheric CO_2 affects the seawater pCO$_2$ and pH. **[2]**

As atmospheric CO_2 increases, the concentration of CO_2 in ALOHA seawater increases as well, while the seawater pH decreases. The evidence shows a positive correlation between atmospheric CO_2 and seawater CO_2 concentrations, and a negative correlation between atmospheric CO_2 concentrations and seawater pH. However, correlation is not causation and other factors could be affecting the pH and ocean CO_2.

ANSWER ANALYSIS

Look at the graph and the trend. It helps to say as (name of X-axis) increases (name of Y-axis) increases or decreases. The question here wants the trend for both pCO2 and pH. Sometimes the graph may reach a peak or plateau – it doesn't in this example. Often it helps if there are error bars to comment on the variation in the data. Here there are no error bars but there is a repeating seasonal trend. It may also be worth commenting that the correlation is not a causation.

ANSWER ANALYSIS

For full marks here, you need to link the theory between carbon dioxide from combustion increasing the pCO$_2$ of the ocean, which decreases pH.

(d) Suggest causes for the changes in ocean acidity. **[3]**

Increased burning/combustion of fossil fuels/deforestation has led to increased CO_2 in the atmosphere.

• The increased greenhouse gases/increased greenhouse effect the more heat/long-wave radiation is trapped in the atmosphere.

• The CO_2 dissolves in water forming (carbonic/H_2CO_3) acid.

• The increase in acid causes the pH to fall.

Acidification of oceans is a threat to marine organisms such as the Antarctic pteropod *Limacina helicina*. The pteropod is a tiny sea snail the size of a pea with a calcium carbonate shell. It is eaten by krill, whales and North Pacific juvenile salmon. A study was carried out in which the pteropod's shell was placed in seawater with the predicted pH (7.8) and carbonate levels projected for 2100. The photos below show the shell transparency over 45 days.

  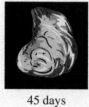

| 0 days | 15 days | 30 days | 45 days |

(e) Identify what phylum the pteropod belongs to. **[1]**

Mollusca

The study below shows the percent mortality of *Limacina helicina* after 29 days of incubation at different water temperatures and pCO_2 concentrations.

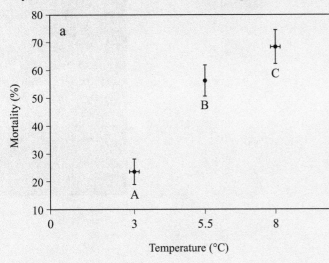

[Source: Lischka, S., Budenbender, J., Boxhammer, T., Riebesell, U., 'Impact of ocean acidification and elevated temperatures on early juveniles of the polar shelled pteropod Limacina helicina: mortality, shell degradation, and shell growth' (*Biogeosciences*, 15 April 2011, CC BY 3.0)]

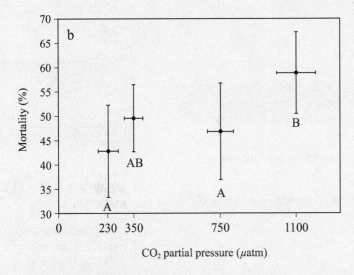

(f) Describe the trend in mortality when temperature is increased. **[2]**

As temperature increases mortality increases and there is a positive correlation between temperature and mortality. There are greater increases between 3°C and 5.5°C than 5.5°C and 8°C.

> Check you are looking at the top graph for temperature!

(g) Explain reasons for the change in the shell of *Limacina helicina* over the 45 days. **[3]**

Carbonic acid (H_2CO_3) dissociates/ionizes. This forms hydrogen ions. The H+ ions react with the carbonate ions. There are less free carbonate ions in the ocean to build shells. Existing shells react with the H+ ions and become thin or dissolve.

> **EXPLAIN**
> The explain question requires a detailed account of the reasons/causes linking acidity to the damage in calcium carbonate shells.

(h) Pteropods are abundant in marine zooplankton. Use this information to predict the impact the increased temperature will have on the salmon fishing industry. **[2]**

The increased temperature will reduce zooplankton which means less food for salmon. The salmon population decreases which will have an economic impact. For example, fishermen/fisherwomen make less money.

Paper 2, Section A, Question 2

The following electron micrograph shows a stem cell.

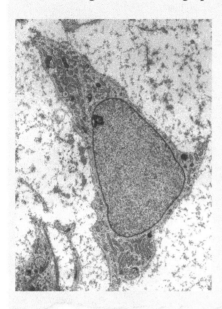

 Remember most of the time the cell is in interphase – if you can see a nuclear membrane clearly the cell is in interphase.

 Sometimes you will need to identify stages of the cell cycle from diagrams.

2. (a) Identify the stage of the cell cycle from the diagram providing **one** observation for your reasoning. [1]

Interphase because the nuclear membrane is visible and the DNA has not condensed.

ANSWER ANALYSIS

You could have also said that stem cells are undifferentiated but have the ability to differentiate. Or that they are multipotent/totipotent/pluripotent.

(b) List **two** characteristics of stem cells. [2]

Stem cells are unspecialized but have the ability to specialize.

Stem cells have the ability for repeated cell division/mitosis.

(c) Explain the control of the cell cycle. [3]

• The cell cycle consists of interphase/G1, S, G2 and mitosis.

• Four cyclins/Cyclins A, D, E, and K regulate the cell cycle.

• The concentrations of cyclins increase and decrease throughout the cycle.

• The change in concentrations of cyclins stimulates the next phase/timing of the cell cycle.

• Cyclins bind to cyclin-dependent kinases to become activated.

• Kinases phosphorylate (target) proteins to activate them.

• Phosphorylated proteins have a particular function in the cell cycle e.g. duplication.

• After the event has happened the cyclins become inactive again.

• Cyclin D stimulates the G1 phase/interphase / Cyclin E stimulates G1 stage to move to S phase / Cyclin A activates DNA

EXPLAIN

This is an explain question so needs a detailed account of how cyclins control the cell cycle.

replication/causes S phase to move to G2 / Cyclin B causes G2 phase to prepare for mitosis/assemble mitotic spindle.

The image below shows 25 cells in different stages of the cell cycle.

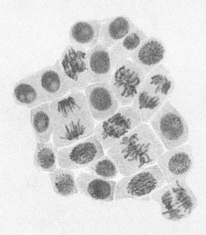

Remember, to calculate a percentage you multiply a fraction by 100.

Here it is:
$$\frac{\text{number of cells in mitosis}}{\text{total cells (given in the question)}} \times 100$$

(d) Suggest which type of microscope was used to produce the image and give one observation for your suggestion. **[1]**

Light microscope as the magnification/resolution is low.

ANSWER ANALYSIS

The examiner would also accept $9 \div 25 \times 100 = 36\%$ or $11 \div 25 \times 100 = 44\%$

(e) Calculate the mitotic index for the cells dividing. **[1]**

$$\frac{10}{25} \times 100 = 40\%$$

3. The bluebell *Hyacinthoides non-scripta* is a native woodland plant endemic to woodland in the United Kingdom. The diagram shows a flower from a hybrid bluebell produced from *Hyacinthoides non-scripta* and the Spanish bluebell, *Hyacinthoides hispanica*.

ANSWER ANALYSIS

Plant phyla include bryophyta, filicinophyta, conipherophyta and angiospermophyta. As there are clearly flowers visible, the phylum is angiospermophyta.

(a) Identify which phylum the bluebell belongs to. **[1]**

Angiospermophyta

(b) Outline the use of the binomial system of nomenclature in categorizing the bluebells. **[2]**

Hyacinthoides non-scripta and Hyacinthoides hispanica: the first name represents the genus and the second is the species.

The unique combination of the two names designates the species worldwide/internationally. They belong to the same genus (Genus Hyacinthoides) but they are different species.

Learn the definitions using flash cards as you can gain easy marks for recall.

4. The table shows the approximate percentages of each base in the genetic material of different species.

Organism	Percentage of each base			
	%A	%G	%C	%T
Octopus	32	18	18	32
Human	30	20	20	30
Grasshopper	29			
Yeast	31	19	19	31
Bacteriophage ΦX174	24	23	22	31

(a) State the percentage of G, C and T in the grasshopper's generic material. **[1]**

%G 21

%C 21

%T 29

> This question is checking you remember which base pairs with which.

(b) Suggest why humans and yeast can be so different despite having similar percentages of bases in their DNA. **[2]**

- They have different genes.
- Bases are in a different order.
- Different base sequences.
- So different amino acids/primary structure of proteins polypeptides.

> **••• SUGGEST**
> You need to develop a possible answer based on the information given and apply your understanding of the topic.

> Think about the sequences, and the effects this has on the proteins.

(c) Use the table to explain the type of DNA the bacteriophage virus has. **[2]**

- Single-stranded DNA.
- DNA as it contains T not U.
- Single-stranded as A isn't the same percent as T / C isn't the same percent as G.
- No complementary base pairing.

> Double-stranded DNA contains all four bases, and the bases are paired up. RNA does not contain thymine but contains uracil.

Paper 2 (Section B) examples

The questions in this paper are sometimes called the essay questions, but they are actually made up of three shorter parts (a), (b) and (c). You will get a choice of three questions and you need to choose **two** of them.

Section B is worth 30 marks, plus an extra 2 marks for quality of communication. That means a potential extra mark for each whole numerical question – it is easy to gain this mark if your answer is in a logical order and easy to follow without the examiner needing to constantly reread the answer. Aim to spend about 50 minutes on this question. Each mark should take approximately 2 minutes.

> Often parts (a), (b) and (c) have a link, and if you can work out the link it may help you to answer the question. For example, part (a) could be to draw an amino acid, part (b) might look at membrane proteins, and then part (c) may look at sickle cell anemia, hormones or enzymes that are made of protein.

Paper 2, Section B, Short-answer questions

First, let's look at some examples of 3–4-mark, short-answer questions in Section B.

5. (a) State four substances transported in blood plasma. **[4]**

- Carbon dioxide
- Urea
- Hormones
- Antibodies/immunoglobulins

6. (a) Draw an exocrine gland cell of the pancreas. **[4]**

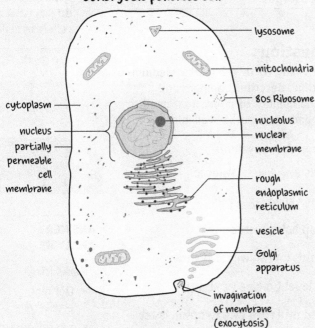

Eukaryotic pancreas cell

6 (b) Draw the structure of a DNA nucleotide. **[3]**

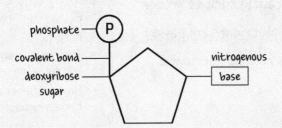

DNA nucleotide

ANSWER ANALYSIS

First, look at (a), (b) and (c) in all question options and look at the highest mark part in each – this is often worth 7 or 8 marks. Check you can answer the 7–8-mark question before choosing which question to answer.

ANSWER ANALYSIS

You could also answer water, glucose or amino acids.

ANSWER ANALYSIS

The drawing correctly has the following:

- Nucleus surrounded by double membrane with pores
- Mitochondria
- Rough endoplasmic reticulum
- Golgi apparatus
- Lysosome
- Plasma membrane, cytoplasm and nucleus
- 80S ribosome/free ribosome

> Don't forget to label your diagrams!

ANSWER ANALYSIS

The drawing correctly has the following:

- Phosphate
- Deoxyribose sugar
- Nitrogenous base / adenine, thymine, cytosine or guanine
- Covalent bond

Now let's look at a **compare and contrast** question in Section B, worth 4 marks.

Compare and contrast questions in Section B are also likely to be short- or medium-answer questions. If you get a compare and contrast question, you need to make sure you put in both the similarities and the differences. You can use a table here or you can use comparative words such as 'greater', 'higher', 'earlier' or 'more'.

COMPARE AND CONTRAST

Describe the similarities and the differences between concepts or items.

Using a table for compare and contract questions really helps as you can make sure the comparisons are there.

5. (b) Compare and contrast eukaryotic and prokaryotic cells. **[4]**

Expect to get 1 mark per comparison.

Eukaryotic cells	Prokaryotic cells
Both have cytoplasm, cell membrane and genetic material	
Nucleus	Nucleoid region
DNA associated with histone proteins	DNA not associated with histone proteins
Linear chromosomes	Circular chromosomes
Membrane-bound organelles (they do have: Golgi apparatus/ rough endoplasmic reticulum/ mitochondria)	No membrane-bound organelles (so no Golgi apparatus / rough endoplasmic reticulum / mitochondria)
80S ribosomes	70S ribosomes

ANSWER ANALYSIS

You could also have mentioned:

- No plasmid → Plasmid
- No pili/flagella → Pili/flagella
- Plant cells have a cellulose cell wall and animal cells do not have a cellulose cell wall

Paper 2, Section B, Medium-answer questions

There are likely to be medium-answer questions in the exam worth 4–5 marks. Medium-answer questions may use command terms such as outline, describe, compare and contrast or explain. Look at the number of marks for the number of points that need to be made and try to make a few more points than the number of marks available.

6. (b) Describe how energy and nutrients move through ecosystems. **[5]**

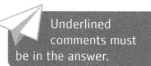
Underlined comments must be in the answer.

Ecosystems are made up of communities and their abiotic environment. Energy enters ecosystems from <u>sunlight/ light energy</u>. Autotrophs/producers <u>trap light energy by photosynthesis</u>. Autotrophs convert light to chemical energy. <u>Energy is then passed from one trophic level to the next by feeding</u>. Only 10% of the energy is passed on from one trophic level to the next – most energy is <u>dissipated as heat as a result of respiration</u>. Energy is lost from the food chain due to feces/inedible bones/inedible parts/hair. <u>Nutrients are constantly cycled through ecosystems</u> (for example, carbon or nitrogen). Nutrients are passed through food chains by eating and saprotrophs recycle nutrients.

ANSWER ANALYSIS

Marks for:

- Defining ecosystems
- Energy enters by sunlight
- How light energy is trapped
- How energy is passed to the next level
- That/how energy is lost between levels
- How nutrients are cycled/examples

6. (c) Outline how the structure of a villus is adapted for its function. **[5]**

Outline requires you to give a brief account or a summary.

Villus

Epithelial Cell

channel protein for facilitated diffusion

protein pump for active transport

many mitochondria (provide energy for active transport)

microvilli (large surface area)

villus wall is one cell thick (short diffusion distance)

lacteal to transport fatty acids

network of capillaries for large surface area for absorption

capillary to remove products of digestion

Don't be afraid to use diagrams if it helps with your answer.

You don't have to use a diagram. You could also write this out as bulleted list like this.

Or you could answer using a table.

- Villus wall/epithelial cell is one cell thick for a short diffusion distance.

- Network of capillaries to increase surface area for absorption.

- Lacteal to absorb fatty acids.

- Epithelial cells have microvilli to increase the surface area for absorption.

- Many protein pumps on epithelial cell membrane for active transport/co-transport of glucose/amino acids.

- Channel proteins on membrane for facilitated diffusion.

- Epithelial cells have many mitochondria for aerobic respiration/ provide energy for active transport.

ANSWER ANALYSIS

Tables are very helpful as they require less writing and ensure both the structure and function are present. It's an efficient way of getting maximum marks.

Paper 2, Section B, Longer-answer questions

The final part of the 'essay question' in Paper 2 (Section B) is the longer-answer question. These questions involve more detail and are often worth 7–8 marks. Keep your answer logical and avoid repetition. We're going to look at three examples here. One is a standard response, one uses graphs, and one uses a Punnett grid in the answer.

It sometimes helps to draft bullet point headings on the side, then put them in order and convert them into sentences.

COMMAND TERMS

These questions are often, but not always, describe, explain, discuss or evaluate questions. 'Explain' requires the causes and reasons, with a detailed account; 'Describe' requires a detailed account; 'Evaluate' requires a discussion of strengths and limitations; 'Discuss' requires a balanced view backed up by evidence.

7. (c) Compare and contrast the structure and use of glycogen and triglycerides in energy storage. [7]

Glycogen is a polysaccharide whereas triglyceride is a lipid. However, both are energy stores in animals. Glycogen is poorly soluble in water and triglycerides are insoluble, so both have little or no osmotic effect.

Triglycerides contain double the energy of carbohydrates – triglycerides provide 37 kJ/g whereas carbohydrates provide 16 kJ/g.

Glycogen is more easily broken down than triglycerides which means it can release its energy more quickly. Glycogen is stored in the muscle and liver whereas triglycerides are stored in adipose tissue.

Glycogen has many 1,4 and 1,6 glycosidic bonds, whereas triglycerides have three ester bonds. Glycogen is hydrolysed into alpha/α-glucose, whereas triglycerides are hydrolysed into one glycerol and three fatty acids. Glucose is a substrate in glycolysis whereas fatty acids are converted to acetyl coenzyme A to enter Krebs cycle.

Carbohydrates/glycogen and amylose are used as short-term energy storage. In contrast, triglycerides are long-term energy storage.

ANSWER ANALYSIS

7 marks means seven points need to be made.

ANSWER ANALYSIS

Remember to write the similarities **and** the differences between concepts or items.

Let's look at another example question, this time with the command term, 'outline'.

7. (c) Outline environmental factors that affect the rate of photosynthesis. [7]

Photosynthesis occurs in chloroplasts of autotrophs and involves using light energy to convert inorganic carbon to organic molecules/glucose. Chlorophyll traps red and blue light energy but reflects green. Environmental factors that affect the rate of photosynthesis include temperature, carbon dioxide and light.

Increasing temperature increases rate of photosynthesis until the optimum temperature is reached. Further increases in temperature decrease photosynthesis as stomata close and enzymes are denatured. Decreasing temperature becomes a limiting factor.

Increasing CO_2 increases photosynthesis until a maximum where it levels off. Decreasing carbon dioxide becomes a limiting factor. Increasing light also increases photosynthesis until a maximum is reached after which it plateaus.

(Answer continued on next page.)

OUTLINE

Give a brief account or a summary.

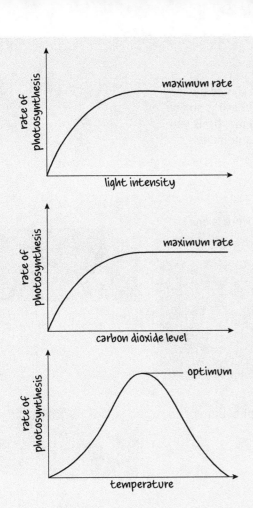

7. (c) Describe how ABO blood types are inherited. **[7]**

There is one gene for blood type. The alleles for blood groups are I^A, I^B and i. The inheritance of blood type is an example of multiple alleles.

Multiple alleles are when there are more than two alleles for a gene.

Alleles I^A and I^B are co-dominant. Allele i is recessive to I^A and I^B.

Phenotype	Genotype
A	$I^A I^A$ and I^A i
B	$I^B I^B$ and I^B i
AB	$I^A I^B$
O	ii

Here's an example.

Heterozygous A and heterozygous B e.g. I^A i x I^B i results in offspring that could be type A, type B, type AB, type O – or another valid cross.

Parent genotype	I^A i		
	Gametes	I^A	i
I^B i	I^B	$I^A I^B$	I^B i
	i	I^Ai	ii

KEY: $I^A I^B$ type AB phenotype / I^A i type A phenotype / I^B i type B phenotype / ii type O phenotype

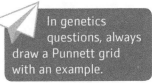

In genetics questions, always draw a Punnett grid with an example.

Paper 3 examples

Paper 3 is 1 hour and 15 minutes long and is worth 45 marks. All answers should be completed in the boxes provided. Section A is compulsory and is worth 15 marks so you should spend a maximum of 25 minutes on it. It is based on the required practicals you have done on the course and can include interpretation of developments or models scientists discovered or proposed or falsified.

Paper 3, Section A, Questions 1–3

Section A of Paper 3 is made up of questions 1–3 and is worth 15 marks. Here is an example of the type of question you will get in Paper 3, Section A.

> Make sure you have reviewed your practical skills. For example, an independent variable is the variable that is changed in an experiment. It is found in the left side heading of a table and on the X-axis of a graph. The dependent variable is the variable that is measured in an experiment. It is found in the right-hand side heading of a table and on the Y-axis of a graph. Controlled variables are parts of the experimental method that are kept the same and may include temperature, volume of a liquid, or mass of a solid, to mention a few.

> Check you are answering the option you learned about in class!

1. The scanning electron micrograph below shows bacteria in the epithelium of stomach lining.

> This is an example of a magnification question linked into digestion.

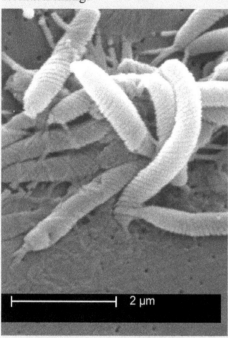

2 µm

(a) Calculate the magnification of the image. **[1]**

Length of scale bar in image = 30 mm = 30,000 µm

Magnification = Image size/Actual size

Magnification = 30,000 / 2

= ×15,000

Paper 3, Section B, Short-answer questions

Let's look at some examples of Section B, short-answer questions. These example questions are taken from Option D. Pay attention to the question structure and the command terms used. There are usually four short-answer questions with three parts, (a), (b) and (c), so the style is similar to your Paper 2 short-answer questions. Then there is a long-answer question.

There are four options — make sure you know which one you have studied. You only need to choose one option out of A, B, C and D and will need to complete all questions in that option.

Option D — Human physiology

1. The energy content of two pieces of dried bread was compared.

   - 25 g of water was placed in a boiling tube
   - Its initial temperature was recorded with a digital thermometer
   - The bread was impaled on a mounted needle
   - The bread was ignited using a Bunsen burner
   - The water was heated using the flame from the burning bread
   - The bread was removed after it had completely burned
   - The final temperature of water was recorded
   - Specific heat capacity of water = 4.2 J °C^{-1} g^{-1}

   The table shows the results of the comparison.

	Wholewheat bread	White bread
Nutritional label values/100 g	620 kJ/100 g, 0.75 grams of fat, 12 grams carbohydrate, 2 grams fibre, 1.5 grams of sugar, 3 grams protein	1151 kJ/100 g, 1 gram of fat, 25 grams carbohydrate, 1 gram fibre, 3 grams of sugar, 4 grams protein
Mass of bread (g)	0.5	0.5
Initial temperature of water (°C)	21.2	21.5
Final temperature of water (°C)	40.1	58.2

   (a) Calculate the energy in kJ/100 g for both breads. **[2]**

	Wholewheat bread	White bread
Change in temperature (°C)	18.9	36.7
Energy released (J)	1984.5	3853.5
Energy released (J/g)	3969	7707
Energy released (kJ/g)	3.969	7.707
Energy released (kJ/100 g)	396.9	770.7

   - Wholewheat bread = 396.9 kJ/100 g
   - White bread = 770.7 kJ/100 g

   (b) Suggest a reason why the results are lower than the expected values of the label? **[1]**

   Not all the energy released is transferred to the water.

   (c) Use the label to suggest why the energy content is higher for the white bread. **[1]**

   White bread has higher carbohydrate and fat.

ANSWER ANALYSIS

You would get one mark if you said wholewheat = 3,969 J/g and white = 7,707 J/g.

ANSWER ANALYSIS

You could also have said that the boiling tube absorbs some of the heat energy. Or that some energy is transferred from the water to the environment.

2. A study in 1996 investigated the relationship between vitamin C dose and the blood plasma vitamin C concentration in seven health volunteers. The previous recommended daily allowance (RDA) was 60 mg. The study investigated oral doses of 30 mg, 60 mg, 100 mg, 200 mg, 400 mg, 1,000 mg and 2500 mg. Error bars represent standard deviation.

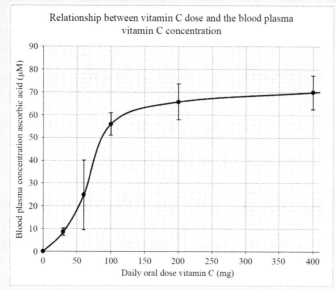

[Source: Data from: Levine, M., et al. Vitamin C pharmacokinetics in healthy volunteers: evidence for a recommended dietary allowance. Proc Natl Acad Sci U S A. 1996 Apr 16;93(8):3704-9. doi: 10.1073/pnas.93.8.3704.]

(a) Vitamin C deficiency is defined as a serum concentration of less than 11.4 µM. State the minimum dose a person should have to prevent deficiency. **[1]**

_40 mg_

(b) Describe the trend between dose of vitamin C and plasma ascorbic acid. **[1]**

As dose increases the plasma ascorbic acid increases until it reaches a plateau.

Sigmoid curve / increases sharply at start then slows down.

Below is a phylogenetic tree of mammals. The highlighted mammals can produce ascorbic acid.

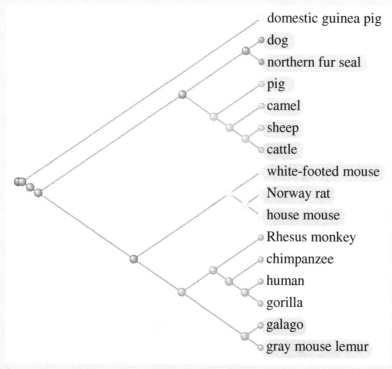

domestic guinea pig
dog
northern fur seal
pig
camel
sheep
cattle
white-footed mouse
Norway rat
house mouse
Rhesus monkey
chimpanzee
human
gorilla
galago
gray mouse lemur

(c) Use the phylogenetic tree to suggest whether the guinea pig or the house mouse should be used as a model organism for vitamin C research. **[1]**

The guinea pig as it is unable to synthesize ascorbic acid (whereas the house mouse is).

The guinea pig as it will show signs of vitamin C deficiency/scurvy whereas the mouse will not.

The images below show a knee joint of a sufferer of scurvy and a sufferer of rickets.

I

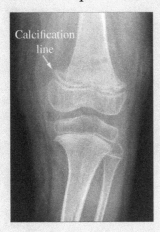

II

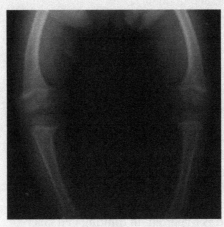

(d) Suggest, with a reason, which image shows rickets. **[1]**

II, as the bones have become bendy.

II, as I shows scurvy with a line of calcification.

(e) State the names of the vitamin and the mineral whose deficiency causes rickets. **[1]**

Vitamin D and calcium.

TESTING WHAT YOU KNOW

In this section, you will be able to test yourself with different sets of practice papers under exam conditions. By taking these mock papers, you will build your confidence and be able to identify any areas you need a bit more practice on. The papers in Set A have a lot of additional guidance in the margin to help you get to the right answer, so attempt this set first.

All you need is this book, a timer, a pen and some extra paper to use if you run out of answer lines. Then you can check your answers at the back of the book when you're done.

Take a deep breath, set your timer, and good luck!

Set A

Paper 1: Higher Level

- Set your timer for **1 hour**
- Each question is worth **[1] mark**
- The maximum mark for this examination paper is **[40 marks]**
- Answer ALL the questions

1. Surface area to volume ratio is important in limiting cell size. Which statement about factors that limit cell size is true? **[1]**
   - ☐ A. A larger cell has a higher rate of metabolism and a larger surface area than a small cell
   - ☐ B. A larger cell has a lower rate of metabolism and a smaller surface area than a small cell
   - ☐ C. A larger cell has a higher rate of metabolism and a smaller surface area to volume ratio than a small cell
   - ☐ D. A larger cell has a lower rate of metabolism and a larger surface area than a small cell

2. This image shows a *Paramecium* magnified under a light microscope × 600. **[1]**

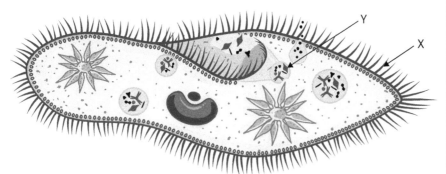

Which letter correctly describes the name and function of X and Y?

		X	Y
☐	A.	Pili involved in reproduction	Nucleoid
☐	B.	Cilia cause movement	Food vacuole
☐	C.	Flagella cause movement	Chloroplast for photosynthesis
☐	D.	Hairs for asexual reproduction	Contractile vacuole for osmoregulation

Margin notes:

You cannot have a calculator for Paper 1.

Transport into a cell speeds up when the cell is smaller as there are more surfaces available per unit volume to let substances diffuse in and out. Small cells have a larger surface area to volume ratio than large cells so diffusion is faster in small cells. Large cells have more metabolism as they have more volume for metabolic pathways.

Paramecium is a single-celled eukaryotic organism. It is a heterotroph.

- **Flagella** are long hair-like organelles that allow movement. There are a few per cell.
- **Cilia** are short hair-like organelles that allow movement in eukaryotes. There are many per cell.
- **Pili** are hair-like structures on the surface of prokaryotes. They are for movement and cell adhesion.
- **Contractile vacuole** for osmoregulation is found at the ends of the *Paramecium*.
- **Chloroplasts** are found in autotrophs.
- **Nucleoid:** the region where a naked loop of DNA is found in prokaryotes.
- **Nucleus:** DNA is enclosed in this in eukaryotes.

3. This image shows striated muscle. Why is striated muscle an exception to cell theory? **[1]**

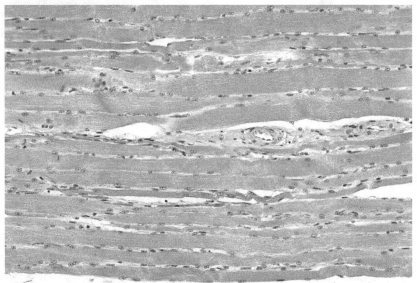

☐ A. Muscle fibres are long and multinucleated
☐ B. The muscle fibres come from pre-existing cells by cell division
☐ C. Muscle fibres are made up of actin and myosin
☐ D. Living organisms are made up of cells

The muscle cells form into long fibres in the embryo, resulting in the cell having many nuclei.

The muscle fibres are up to 300 mm long so are larger than normal cells.

The three principles of cell theory:
• All organisms are made of cells
• Cells are the smallest **unit** of life
• Cells come from pre-existing cells.

4. The electron micrograph below shows a eukaryotic organelle. **[1]**

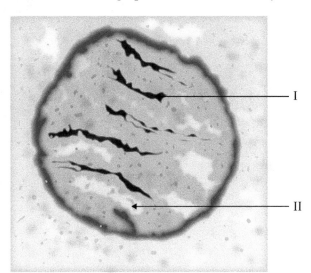

What is the name of the organelle and what are the locations I and II?

		Name	I	II
☐	A.	Chloroplast	Thylakoid Membrane	Stroma
☐	B.	Mitochondrion	Cristae	Matrix
☐	C.	Chloroplast	Cristae	Matrix
☐	D.	Mitochondrion	Thylakoid membrane	Stroma

Mitochondria have a folded inner membrane forming cristae, where the electron transport chain in respiration is found. Mitochondria do not contain thylakoids.

A chloroplast inner membrane is not folded. Chloroplasts contain thylakoids, which are internal membrane-bound compartments where the light-dependent reactions of photosynthesis occur.

The gel-like substance in the mitochondria is called the matrix (site of Krebs cycle) whereas in a chloroplast it is called the stroma (site of Calvin cycle).

5. Which of the following is **not** evidence for endosymbiotic theory? **[1]**

☐ A. Chloroplasts and mitochondria have a naked loop of DNA

☐ B. Chloroplasts and mitochondria are surrounded by a double membrane

☐ C. Chloroplasts and mitochondria are the same size as prokaryotes

☐ D. Chloroplasts and mitochondria have 80S ribosomes

If the ancestor of mitochondria and chloroplasts were eukaryotes, they would also have 80S DNA.

Endosymbiotic theory assumes a larger host prokaryote engulfed but did not digest a smaller aerobic prokaryote. They lived in a symbiotic relationship where the small prokaryote provided energy and the large prokaryote provided food. It is thought this is the ancestor of eukaryotic cells. A similar process later engulfed photosynthetic bacteria which became the ancestor of chloroplasts. Prokaryotes, mitochondria and chloroplasts have 70S ribosomes whereas eukaryotes have 80S ribosomes which supports the theory that mitochondria and chloroplasts are derived from the engulfed smaller prokaryotes. The double membrane of chloroplasts and mitochondria is derived from the cell membrane of the host prokaryote (outer membrane) and engulfed prokaryote (inner membrane).

Cholesterol is a lipid so it contains a high proportion of CH_3.

Cholesterol is a steroid so it has four hydrocarbon rings.

6. The image below shows cholesterol. **[1]**

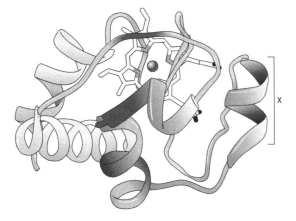

What evidence in the image suggests cholesterol is a steroid?

☐ A. It has many CH_3 groups and four carbon rings

☐ B. It is made up of four glucose rings

☐ C. It has a phosphate head and two hydrocarbon chains

☐ D. It is made up of amino acids

The primary structure of proteins is the number and sequence of amino acids.

The secondary structure of proteins is how the primary structure is held together by hydrogen bonds between the amino and carboxyl groups of non-adjacent amino acids. It includes:

• Beta pleated sheet: the primary structure is folded into a zig-zag shape – in IB exams arrows emphasize the zig zag

• Alpha helix: the primary structure is coiled into a helical (coiled) shape.

7. The image below shows the protein cytochrome c, a protein found in the electron transport chain in mitochondria. What is X? **[1]**

☐ A. Primary structure

☐ B. Alpha helix

☐ C. Beta pleated sheet

☐ D. Heme prosthetic group

The tertiary structure of proteins is the 3D conformation (shape) of the protein held together by interactions between the R groups. This includes ionic bonds, hydrogen bonds, hydrophobic interactions and disulfide bridges.

• **Tertiary structure** is fibrous (long and thin) or globular (compact and rounded). Cytochrome c has a globular tertiary structure.

• **Quaternary structure** is when there is more than one polypeptide chain. Cytochrome c is only one polypeptide so doesn't have quaternary structure.

 A conjugated protein contains a prosthetic group that is not made up of amino acids. Cytochrome c is a conjugated protein.

 X has a helical shape, so which type of secondary structure does that mean it has?

8. *Diplocarpon rosae* is a fungus that infects leaves from roses causing black spot disease. Select which statement could be true of a plane with black spot disease. **[1]**

	Light absorbed	**Oxygen released**
☐ A.	More absorbed	More released
☐ B.	Less absorbed	Less released
☐ C.	More absorbed	Less released
☐ D.	Less absorbed	More released

Photosynthesis uses carbon dioxide and produces oxygen.

carbon dioxide	+	water	→	glucose	+	oxygen
$6CO_2$	+	$6H_2O$	→	$C_6H_{12}O_6$	+	$6O_2$

The black spot could result in less chlorophyll and less absorption of light. As a result, there would be less photosynthesis so less oxygen released.

9. What is a nucleosome? **[1]**
   ☐ A. Naked DNA
   ☐ B. DNA associated with histone proteins
   ☐ C. Deoxyribose sugar, base and phosphate
   ☐ D. Ribose sugar, base and phosphate

Naked DNA is found in prokaryotes and is DNA not associated with histone proteins.

A nucleosome is made up of a section of DNA strand wrapped twice around eight histone proteins (octamer). Many nucleosomes are linked by an H1 histone protein and coiled and compressed to form chromatin.

DNA and RNA strands are made up of nucleotides.
- The nucleotide in **DNA** is made up of **deoxyribose** sugar, base and phosphate.
- The nucleotide in **RNA** is made up of **ribose** sugar, base and phosphate.

10. If 30% of DNA is cytosine how much is adenine? **[1]**
    ☐ A. 70%
    ☐ B. 50%
    ☐ C. 20%
    ☐ D. 10%

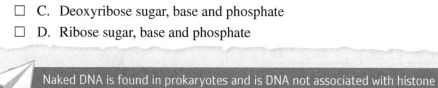

Remember that **(A+T) + (C+G) = 100%** because DNA is double stranded.

If there is 30% cytosine in the DNA then how much must be guanine? Think of complementary base pairing (C+G). The remaining DNA is the percentage for both adenine and thymine (A+T). How would you find adenine only?

11. Below is an image of a DNA nucleotide. **[1]**

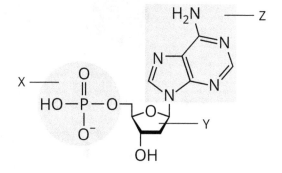

- A–T pairs have three hydrogen bonds.
- C–G pairs have three hydrogen bonds.
- A and G are purines and have two nitrogenous rings in their structure.
- C and T have one nitrogenous ring in their structure so are pyrimidines.

Identify: X, Y and Z.

		X	Y	Z
□	A.	Phosphorus	Ribose sugar	Adenine
□	B.	Phosphate	Deoxyribose sugar	Thymine
□	C.	Phosphate	Deoxyribose sugar	Adenine
□	D.	Phosphorus	Ribose sugar	Thymine

• Z has two rings and (two hydrogen bonds) and therefore must be A (adenine).

12. Which sequence of bases and amino acids could be made by the transcription and translation of the DNA molecule shown? **[1]**

3' ATGGAATATCGATTTAAA 5'

5' TACCTTATAGCTAAATTT 3'

The table uses mRNA codons so you need to work out the mRNA sequence.

		2nd base in codon					
		U	C	A	G		
1st base in codon	U	Phe	Ser	Tyr	Cys	U	3rd base in codon
		Phe	Ser	Tyr	Cys	C	
		Leu	Ser	Stop	Stop	A	
		Leu	Ser	Stop	Trp	G	
	C	Leu	Pro	His	Arg	U	
		Leu	Pro	His	Arg	C	
		Leu	Pro	Gin	Arg	A	
		Leu	Pro	Gin	Arg	G	
	A	Ile	Thr	Asn	Ser	U	
		Ile	Thr	Asn	Ser	C	
		Ile	Thr	Lys	Arg	A	
		Ile	Thr	Lys	Arg	G	
	G	Val	Ala	Asp	Gly	U	
		Val	Ala	Asp	Gly	C	
		Val	Ala	Glu	Gly	A	
		Val	Ala	Glu	Gly	G	

The DNA code containing the genetic information is on the **sense strand**, which is read in the 5' to 3' direction. The **antisense strand** is used as the template to make the mRNA. The mRNA will read the same as the DNA 5' to 3' but with a U instead of T. mRNA contains U instead of T and matches the sense strand from the 5' carbon to 3' carbon direction.

mRNA = UACCUUAUAGCUA AAUUU

• Look up UAC in the table: U = First row, A = third column, C = second line.
• Read the amino acid name in the first row, third column, second line = Tyr.
• Repeat with the rest of the codons.

		Sequence of bases	Sequence of amino acids
□	A.	UAC-AUA-AAA-UUU-CUU-GCU	Tyr-Ile-Lys-Phe-Leu-Ala
□	B.	UAC-CUU-AUA-GCU-AAA-UUU	Tyr-Leu-Ile-Ala-Lys-Phe
□	C.	UAU-AUC-GAA-CGA-UUU-AAA	Tyr-Ile-Glu-Arg-Phe-Lys
□	D.	UAU-GAA-CGA-UUU-AAA	Tyr-Glu-Arg-Phe-Lys

13. What is the advantage of mRNA splicing in eukaryotes? **[1]**

- □ A. Exons are removed
- □ B. It allows DNA replication
- □ C. Many ribosomes can translate a protein at the same time
- □ D. A single gene can code for multiple proteins

mRNA splicing is when non-coding regions of a gene called introns are removed from immature mRNA to make mature mRNA. Only the exons are then expressed in the polypeptide after translation.

Different tissues can splice the same mRNA in different ways leading to different mature mRNA and therefore different polypeptides being made from the same gene.

DNA replication refers to how DNA copies itself by semi-conservative replication.

14. What is a polysome? **[1]**

- ☐ A. DNA and histone protein
- ☐ B. Multiple ribosomes attached to one mRNA molecule
- ☐ C. Pentose sugar, nitrogenous base and phosphate
- ☐ D. Naked loop of DNA

Polysomes are many ribosomes translating one mRNA at a time. All the polypeptides made are identical.

15. The diagram below represents the Krebs cycle.

The nucleotide in DNA is made up of deoxyribose sugar, nitrogenous base and phosphate.

Decarboxylation is the removal of carbon dioxide from a molecule.

Reduction is the addition of hydrogen, addition of electrons or removal of oxygen.

Oxidation is the loss of hydrogen, removal of electrons or addition of oxygen.

Which processes are correct? **[1]**

☐ A.	C_6, C_5 and C_4 are reduced	C_6 and C_5 are carboxylated	NAD is oxidized
☐ B.	C_6, C_5 and C_4 are oxidized	C_6 and C_5 are carboxylated	NAD is oxidized
☐ C.	C_6, C_5 and C_4 are oxidized	C_6 and C_5 are decarboxylated	NAD is reduced
☐ D.	C_6, C_5 and C_4 are reduced	C_6 and C_5 are decarboxylated	NAD is reduced

16. What is chemiosmosis? **[1]**

- ☐ A. Generating ATP as a result of the movement of protons across a membrane
- ☐ B. The movement of chemicals from a dilute to a concentrated solution
- ☐ C. The movement of water across a partially permeable membrane
- ☐ D. The selective reabsorption of glucose in the proximal convoluted tubule

- **Diffusion** is the movement of particles from a high to low concentration.
- **Facilitated diffusion** is the movement of particles from high to low concentration down the concentration gradient through a channel protein.
- **Osmosis** is the movement of water from a dilute (hypotonic) to a concentrated (hypertonic) solution across a partially permeable membrane.

17. The diagram below shows a eukaryotic organelle. What is the name of X and what reaction is occurring there? **[1]**

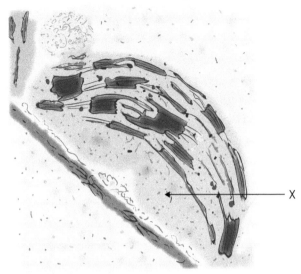

	X	Name of reaction
☐ A.	Matrix	Phosphorylation
☐ B.	Stroma	Photolysis
☐ C.	Matrix	Krebs cycle
☐ D.	Stroma	Calvin cycle

18. A mouse has the diploid number of 40 chromosomes. How many autosomes are in a mouse sperm cell? **[1]**

☐ A. 20

☐ B. 19

☐ C. 40

☐ D. 38

Autosomes are any chromosomes that do not determine the sex of the mouse. There are 38 autosomes (non-sex cells) and 2 sex cells (X and Y) out of the 40.

The diploid number of chromosomes in a mouse is 40 (or 20 pairs). When gametes are made only one of each chromosome pair will pass into the gamete.

The sperm cell will have the haploid number of 20 chromosomes. One of which will be a sex cell so that leaves 19 autosomes.

19. What possible phenotypes could the offspring of a parent who is heterozygous for blood group A and a second parent heterozygous for blood group B have? **[1]**

☐ A. AB only

☐ B. A, B

☐ C. A, B, AB, O

☐ D. A only

20. The diagram shows a pedigree of polydactyly, in which an affected person has an extra finger. Individuals with the disorder are shown as shaded.

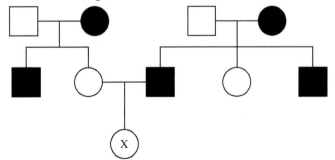

Recessive traits will have some unaffected parents having affected children. It is not possible for two affected parents (homozygous recessive) to have an unaffected child. **Dominant** traits will have a parent affected in every generation. It is possible for two affected parents to have an unaffected child if both parents are heterozygous.

What is the chance that the individual labelled X has the disorder? **[1]**

☐ A. 100%

☐ B. 50%

☐ C. 25%

☐ D. 0%

You can work out that polydactyly is an **X-linked recessive** trait.

21. In the fruit fly *Drosophila melanogaster* the genes for body colour and wing type are linked. The allele for ebony body colour, allele (e), is recessive to the yellow wild-type body colour, allele (E). The allele for vestigial (vg) wings is recessive to the longer wild-type wing (Vg)

A test cross resulted in these recombinants:

$$\frac{e \quad Vg}{e \quad vg} \qquad \frac{E \quad vg}{e \quad vg}$$

Which of the following shows the genotypes of both parents? **[1]**

☐ A. $\dfrac{E \quad Vg}{e \quad vg} \times \dfrac{e \quad vg}{E \quad vg}$

☐ B. $\dfrac{E \quad Vg}{e \quad vg} \times \dfrac{E \quad Vg}{e \quad vg}$

☐ C. $\dfrac{E \quad Vg}{e \quad vg} \times \dfrac{e \quad vg}{e \quad vg}$

☐ D. $\dfrac{E \quad Vg}{e \quad Vg} \times \dfrac{e \quad vg}{e \quad vg}$

Treat Vg and vg each as a single letter inherited together – never split the V from the g (vg) or (Vg). A test cross means crossing with a **homozygous** recessive. This means A and B can be eliminated.

As one chromosome must come from each parent, the offspring must have a chromosome from the **recessive** parent in the test cross. This means A and D can be eliminated.

ANSWER ANALYSIS

C is correct as the offspring have inherited the e vg chromosome from the test cross. The chromosomes in the question are recombinants; therefore, the parent must have a different combination from e Vg or E vg so they must have had E Vg and e vg.

22. The image shows chromosomes in meiosis. What stage of meiosis are they in? **[1]**

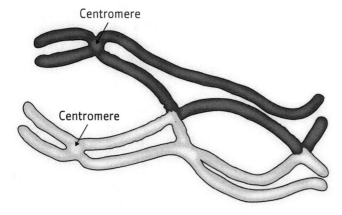

Centromere

Centromere

☐ A. Prophase I

☐ B. Prophase II

☐ C. Anaphase I

☐ D. Anaphase II

The overlaps where homologous chromosomes cross over are called chiasmata and are a source of variation in gametes. This crossing over occurs in prophase I of meiosis between non-sister chromatids.

To count the number of chiasmata, count the number of overlaps between the chromosomes. The centromeres of each chromosome are labelled. Count the number of centromeres to find the number of chromosomes.

23. What type of enzyme could cut a DNA molecule as shown in the diagram below? **[1]**

```
G A G┆T T A A T T C T C
      ┆
      └ ─ ─ ─ ─ ─ ─ ┐
                    ┆
C T G A A T T A A┆G A G
```

- ☐ A. DNA gyrase
- ☐ B. DNA ligase
- ☐ C. RNA polymerase
- ☐ D. Restriction enzyme

> DNA **gyrase** relieves the tension in the supercoiled DNA in semi-conservative replication.

> DNA **ligase** covalently bonds the sugar–phosphate backbone on Okazaki fragments in semi-conservative replication.

> RNA **polymerase** breaks the hydrogen bonds in DNA during transcription.

> A restriction enzyme cuts DNA at a specific point. It is useful in DNA profiling and in making recombinant DNA for genetic engineering e.g. opening the bacterial plasmid for the insertion of the insulin gene.

24. The graphs below shows the mean monthly carbon dioxide measured at Mauna in Hawaii.

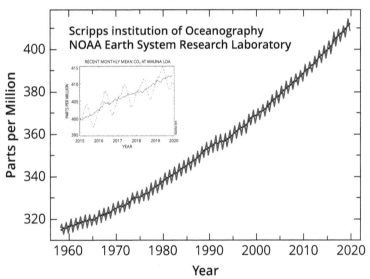

[© NOAA (Data: Mauna Loa (ftp://aftp.cmdl.noaa.gove/products/trends/co2/co2_mm_mlo.txt) ALOHA (http://soest.hawaii.edu))]

Which of the following conclusions can be made from the data in the graph?

I.   The level of CO_2 is increasing between 1960 and 2020
II. There are seasonal variations in the data
III. Global warming causes CO_2 levels to change **[1]**

- ☐ A. I only
- ☐ B. II only
- ☐ C. I and II
- ☐ D. II and III

> There is a positive correlation, but a correlation does not prove causation.

> The fluctuation occurs due to changes in the rate of photosynthesis. Most land mass is in the northern hemisphere. In summer in the northern hemisphere there is more photosynthesis, so CO_2 levels fall. In winter there is less photosynthesis, so CO_2 levels rise.

25. Which conditions are required for the formation of peat? **[1]**

☐	A.	Organic matter	Acidic conditions	Anaerobic conditions
☐	B.	Inorganic matter	Alkaline conditions	Aerobic conditions
☐	C.	Organic matter	Alkaline conditions	Aerobic conditions
☐	D.	Inorganic matter	Acidic conditions	Anaerobic conditions

> Peat requires the pH to be low and a lack of oxygen. If oxygen is present, aerobic decomposers will break down the organic matter.

26. What are the three domains in the hierarchy of taxa? **[1]**

- ☐ A. Angiospermophytes, Coniferophyte, Bryophyte
- ☐ B. Cnidaria, Orthropoda, Olathyhellminths
- ☐ C. Fungai, Animalia, Olantae
- ☐ D. Eubacteria, Archaea, Eukarya

Binomial names used by Linnaeus are the genius and species names.
- **Species** – group of similar organisms that can interbreed to make fertile offspring.
- **Genus** – group of similar species.
- **Kingdom** – includes Animalia, Plantae, Fungi, Protista.

Use this mnemonic to remember the hierarchy of taxa:

Do	Keep	Pond	Clean	Or	Frog	Gets	Sick
Domain	Kingdom	Phylum	Class	Order	Family	*Genus*	*Species*

Phylum are groups of similar classes:
- **Animal phyla** include Cnidaria, Arthropoda, Platyhelminthes.
- **Plant phyla** include Angiospermophytes, Coniferophytes, Bryophytes.

27. Which statement shows some of the characteristics of Cnidaria? **[1]**

- ☐ A. Bilateral symmetry, mouth and no anus
- ☐ B. Radial symmetry, may have stinging cells
- ☐ C. Jointed appendages with exoskeleton
- ☐ D. Muscular foot and mantle

Domains:
- **Archaea:** extremophile bacteria (e.g. thermophiles and methanogens).
- **Eubacteria:** rest of common bacteria.
- **Eukarya:** have a nucleus (e.g. Protista such as *Paramecium*; Fungi such as *Ryhtisma acerinum*; Animalia such as *Gorilla gorilla*; Plantae such as *Solanum tuberosum*).

Arthropoda have bilateral symmetry, a mouth and an anus, jointed appendages, and an exoskeleton (e.g. scorpion).

Platyhelminthes have bilateral symmetry, no mouth or anus and are flat (e.g. tapeworms).

Cnidaria have radial symmetry, a mouth, no anus and tentacles with stinging cells (e.g. jellyfish)

28. The protein cytochrome c is found in mitochondria in the electron transport chain. This makes it useful to compare the DNA and amino acid sequences of organisms.

Below are some amino acid sequences from cytochrome c in four organisms. The differences in sequences between organisms are in **bold**.

Which cladogram correctly matches the data? **[1]**

The more amino acids a species has in common, the more closely related the species are.

The species with the fewest amino acids in common is the earliest on the cladogram.

Organism	Amino acid sequence from cytochrome c												
dolphin	LEU	ISO	PRO	PRO	PHE	ILE	LEU	LEU	SER	**HIS**	**VAL**	**VAL**	**SER**
cat	LEU	ISO	PRO	PRO	PHE	ILE	LEU	LEU	SER	**HIS**	LEU	LEU	**SER**
sponge	**ISO**	ISO	**ASP**	**GLN**	PHE	ILE	LEU	**HIS**	SER	ARG	LEU	LEU	ARG
shark	LEU	ISO	PRO	PRO	PHE	ILE	LEU	LEU	SER	ARG	LEU	LEU	ARG
earthworm	LEU	ISO	**ASP**	PRO	PHE	ILE	LEU	**HIS**	SER	ARG	LEU	LEU	ARG

Draw a table to show the number of similarities in amino acid sequences. You can use the data from either the dolphin or the sponge to draw a cladogram as the data for all species differs. Note that dolphin and sponge have the fewest number of similarities so should be the most distantly related. Use either the sponge or dolphin to rank the number of similarities.

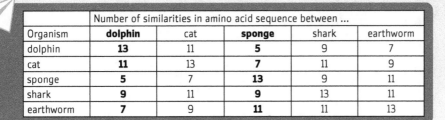

	Number of similarities in amino acid sequence between ...				
Organism	**dolphin**	cat	**sponge**	shark	earthworm
dolphin	**13**	11	**5**	9	7
cat	**11**	13	**7**	11	9
sponge	**5**	7	**13**	9	11
shark	**9**	11	**9**	13	11
earthworm	**7**	9	**11**	11	13

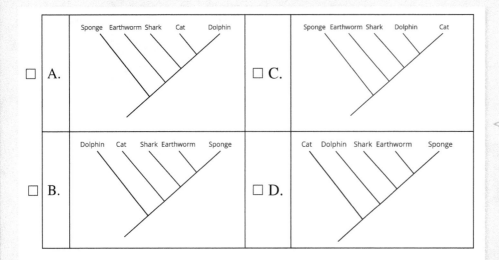

□ A.

□ C.

□ B.

□ D.

29. How do greenhouse gases affect global warming? **[1]**

    □ A. They trap shortwave radiation arriving at the Earth from the sun

    □ B. They trap longwave radiation re-radiated from the Earth's surface

    □ C. They trap longwave radiation arriving at the Earth from the sun

    □ D. They trap shortwave radiation re-radiated from the Earth's surface

30. In the graph below, the X-axis represents the phenotypic trait and the Y-axis represents the number of organisms. What type of natural selection does this represent? **[1]**

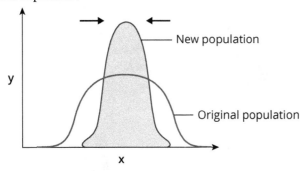

    □ A. Directional natural selection

    □ B. Stabilizing natural selection

    □ C. Disruptive natural selection

    □ D. Diverging natural selection

> The three types of natural selection are directional, stabilizing and disruptive selection.

> Stabilizing selection occurs if the intermediate phenotype is favoured. This has occurred in the birthweight of humans: fewer very low and very high birthweight babies survive to reproduce, meaning the alleles for extremely high and low birthweights become less common.

> Disruptive selection is where the extreme phenotypes are favoured over the intermediate. An example of this is the peppered moth, which has light or dark variants but few intermediate types. Disruptive selection is an evolutionary force driving a population apart and can lead to reproductive isolation and speciation.

> • Shorter wavelengths (e.g. light and ultraviolet light) are emitted from the sun.
> • They are absorbed by the surface of the Earth and re-emitted as longer wavelength infrared radiation (heat).
> • The longer wavelength radiation cannot escape the Earth's atmosphere as easily as the shorter wave radiation, so is reflected back to Earth.

> The original population is favouring a single phenotype.

> Natural selection occurs when there is variation in a population.

> A change in the environment makes some organisms better adapted than others. This selection pressure leads to the survival of the fittest organisms who reproduce more and pass on their alleles to their offspring. This means that over time the gene pool changes to favour the fitter alleles, resulting in a change in the phenotype.

> Directional selection occurs if an extreme phenotype is favoured. An example of this is Darwin's finches on the Galápagos Islands, which show directional selection adapted for their specific island. For example, the large ground finch (*Geospiza magnirostris*) has developed a large, short beak for cracking nuts.

31. Which statement about evolution is correct? **[1]**

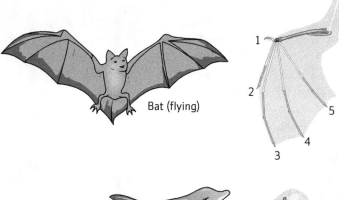

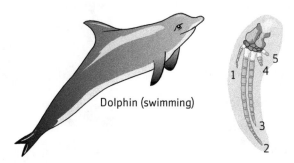

Bat (flying)

Dolphin (swimming)

☐ A. The bat wing and dolphin flipper are homologous in structure, giving evidence for convergent evolution

☐ B. The bat wing and dolphin flipper are homologous in structure, giving evidence for divergent evolution

☐ C. The bat wing and dolphin flipper are analogous in structure, giving evidence for convergent evolution

☐ D. The bat wing and dolphin flipper are analogous in structure, giving evidence for divergent evolution

Homologous evolution refers to similar structures that have adapted over time to develop different functions (adaptive radiation) as a result of natural selection. This similarity in structure suggests the organisms have a common ancestor.

Analogous structures are not similar in their structure but do have a similar function.

32. Which structure is shown in the following image? **[1]**

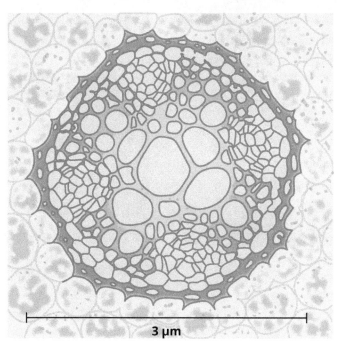

3 μm

☐ A. Cross-section of a leaf

☐ B. Cross-section of a dicotyledon stem

☐ C. Cross-section of a flower

☐ D. Cross-section of a dicotyledon root

The xylem is shown in the X-shaped centre. Xylem is made up of dead cells with bands of lignin for strength.

The phloem is shown around the outside of the xylem and transports sugars from source by translocation. Phloem contains living sieve tube cells and companion cells.

33. What is the correct sequence for the germination of a broad bean (*Vicia faba*) seed? **[1]**

☐ A.	Imbibes water	Gibberellin produced	Amylase breaks down starch to maltose	Plumule emerges followed by radicle
☐ B.	Gibberellin produced	Imbibes water	Amylase breaks down starch to maltose	Radicle emerges followed by plumule
☐ C.	Imbibes water	Gibberellin produced	Amylase breaks down starch to maltose	Radicle emerges followed by plumule
☐ D.	Gibberellin produced	Imbibes water	Amylase breaks down starch to maltose	Plumule emerges followed by radicle

34. Which structure in the kidney is correctly matched to its function? **[1]**

	Structure	Function
☐ A.	Glomerulus	Selective reabsorption
☐ B.	Proximal convoluted tubule	Ultrafiltration
☐ C.	Loop of Henle	Creates a low solute concentration in the medulla
☐ D.	Collecting duct	Osmoregulation

The loop of Henle creates a high Na+ concentration in the medulla to reabsorb water from the filtrate.

The collecting duct alters the concentration of the urine. ADH makes the collecting duct membrane more permeable to water (more aquaporins) so more water is reabsorbed into blood capillaries and a smaller volume of concentrated urine is made.

35. What is the correct sequence of blood clotting? **[1]**
- ☐ A. Prothrombin – thrombin – fibrinogen – fibrin
- ☐ B. Fibrinogen – fibrin – prothrombin – thrombin
- ☐ C. Fibrin – fibrinogen – prothrombin – thrombin
- ☐ D. Thrombin – prothrombin – fibrin – fibrinogen

Clotting factors released by the damaged cell cause platelets to plug the damaged area.

A seed needs oxygen, water and warmth to germinate. In these conditions it will absorb water through the micropyle. This rehydrates the tissue allowing the embryo to make the hormone gibberellin. The seed then produces the enzyme amylase which breaks starch to maltose. Maltose is then broken down to glucose for respiration or used to make cellulose for the cell wall. Energy from respiration is used to build tissue that grows into a radicle (small root) and then a plumule (small shoot) followed by photosynthetic leaves.

The glomerulus is a ball of capillaries inside the Bowman's capsule. Small molecules are filtered from the blood in the capillaries into Bowman's capsule by ultrafiltration.

The proximal convoluted tubule under the Bowman's capsule is where glucose and water and Na+ are selectively reabsorbed back into blood capillaries.

Clotting factors also cause the soluble globular protein prothrombin to be converted to the active globular protein – the enzyme thrombin. Thrombin then catalyses the conversion of the soluble globular protein fibrinogen into the insoluble protein fibrin. Fibrin than traps red blood cells and the platelets. This forms a scab that acts as a barrier to pathogens.

36. Which statement about alveoli is correct? **[1]**

☐ A. Type I pneumocytes contain surfactant to reduce surface tension in alveoli

☐ B. Type II pneumocytes are very thin to allow gas exchange

☐ C. Oxygen diffuses from alveoli into blood capillaries and carbon dioxide diffuses into alveoli

☐ D. Oxygen diffuses out of blood capillaries and oxygen diffuses in

> Type I alveoli cells (pneumatocytes) are thin and allow gas exchange due to their short diffusion distance. This allows fast diffusion of oxygen and carbon dioxide.

> Gas exchange occurs in the alveoli as the type I pneumatocytes are thin.

> Oxygen diffuses from the higher concentration in the alveoli to the lower concentration in the blood capillaries.

> Carbon dioxide diffuses from the higher concentration in the blood capillaries to the lower concentration in the alveoli.

> Type II alveoli cells (pneumatocytes) make a surfactant which decreases the surface tension and stops the alveoli walls from sticking together.

37. The graph shows stages in the propagation of a nerve impulse. Which letter shows repolarization? **[1]**

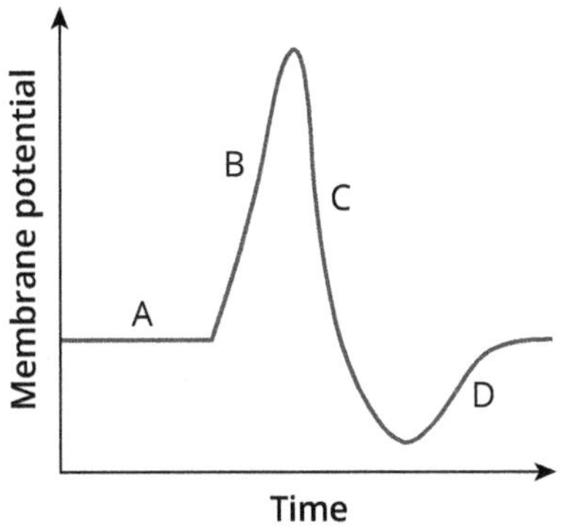

☐ A. A

☐ B. B

☐ C. C

☐ D. D

> The **resting potential** of neuron is −70 mV, meaning the inside of a neuron is more negative than the outside. The neuron is not sending an electrical impulse.

> If an **action potential** arrives, sodium voltage-gated channels open and sodium ions diffuse into the axon. If the threshold potential (−50 mV) is reached, an influx of sodium ions causes the inside of the neuron to be positive compared with the outside. The polarity is now positive not negative, so we say the neuron is depolarized. This positive charge will cause a localized current to flow.

> **Repolarization**: the change in polarity causes potassium voltage-gated channels to open and potassium ions diffuse out of the axon (−80 mV).

> **Refractory period**: active transport pumps move two potassium ions back into the axon for every three sodium ions it pumps out, restoring the resting potential.

38. Where is the sinoatrial node in the heart located? **[1]**

☐ A. Left atrium

☐ B. Left ventricle

☐ C. Right atrium

☐ D. Right ventricle

> The sinoatrial node (or pacemaker) is a group of nerve tissue that sends out electrical impulses to coordinate the myogenic heart muscle cells to contract at the same time.

39. What is the function of ATP in muscle contraction?

I.   To bind to troponin, exposing the myosin head binding sites
II. To break the cross-bridges between actin and myosin
III. To be hydrolysed into ADP + Pi to give energy for the power stroke   **[1]**

☐ A. I only

☐ B. I and II only

☐ C. II and III only

☐ D. I, II and III

> There are two types of muscle fibres in the muscle cell (sarcomere). Both fibres are proteins with a fibrous tertiary structure. **Actin** is the thin fibre and **myosin** is the thick fibre covered in myosin heads.

> ATP binds to the myosin head, breaking the cross-bridges that formed between actin and myosin heads. The ATP is then broken down into ADP + Pi, releasing energy. This energy is used to cock the myosin head by moving it on to an actin binding site further from the centre of the sarcomere. When ADP and Pi are released, the myosin head returns to its uncocked shape. This change in shape slides the actin over the myosin, shortening the sarcomere.

> An action potential causes calcium ions to be released from the sarcoplasmic reticulum. The actin protein has binding sites for the myosin protein heads; however, these binding sites are blocked by tropomyosin. Calcium ions bind to troponin, which causes tropomyosin to alter its position on the actin, exposing the myosin head binding sites.

40. The images below are of the same muscle fibre.
Compare the images and deduce which statement is correct.   **[1]**

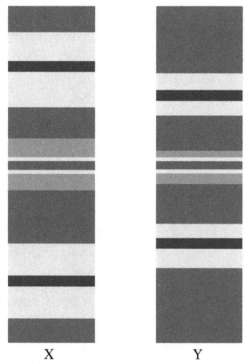

X   Y

> The two very dark lines are the Z lines. The distance between the two Z lines is the length of the sarcomere. The distance between the Z lines is smaller in Y.

> The light bands either side of the Z line are actin filaments only (I band). The light bands at the ends of the sarcomere are wider in X than Y so there is more actin not overlapping.

☐ A. X is contracted and the overlap between actin and myosin has increased

☐ B. X is relaxed and the overlap between actin and myosin has increased

☐ C. Y is contracted and the overlap between actin and myosin has increased

☐ D. Y is relaxed and the overlap between actin and myosin has increased

> The dark band in the centre of the sarcomere is the region of overlap between the actin and myosin. The overlap is greater in Y.

Paper 2: Higher Level 🔢

- Set your timer for **2 hours and 15 minutes**
- The maximum mark for this examination paper is **[72 marks]**
- **Section A** – answer ALL the questions
- **Section B** – answer **two** questions
- A calculator is needed for this paper

Section A

1. Bovine tuberculosis (TB) is a disease that can be transferred from wildlife hosts to mammalian species. The infection of livestock (such as cows – *Bos taurus*) results in severe economic losses from death and disease.

   TB is caused by the bacteria *Mycobacterium bovis* and is spread from wild animals such as the Australian brushtail possum (*Trichosurus vulpecula*). TB can be transmitted to humans as well as to other animals.

   (a) State the domain the infected cows belong to. **[1]**

   ....

The paragraph at the top of Q1 tells you some basic information about the topic. Annotate in the margin the area of the syllabus that the paragraph loosely relates too. This will help you later. It may link more than one area – for example, human biology, diabetes and stem cells. This will focus your mind for the later parts of Q1.

ANSWER ANALYSIS

The first question is a straightforward question such as reading off a graph or a **state** question in which you need to recall/read off the one correct answer. The domain it belongs to is Eukaryota. You don't need to justify your answer at all. Just a one-word answer will get you the mark here. If it is a read off the graph question then check you are reading off the correct line/bar/scale.

!

Don't get confused with the genus. The genus the cows belong to is Bos.

ANSWER ANALYSIS

The phrase 'transmitted to humans' links with defence against disease. The key words of 'disease' and 'bacteria' also do this.

   (b) Use information from the text above to identify the treatment for humans who have contracted bovine tuberculosis. **[1]**

   ....

   ....

   A study in New Zealand investigated if the human BCG vaccination was effective in preventing TB in wild possums. An oral vaccine in food was delivered to one group of wild possums. A second control group did not receive the vaccine. Initial vaccinations were applied in November–December 2004 and the animals were tagged and released back into the wild. Every two months for two years the scientists recaptured the animals and tested them for the presence of TB. The vaccine was reapplied to recaptured possums at six-month intervals.

ANSWER ANALYSIS

You might consider vaccinations; however, vaccination would not be appropriate as the person has already contracted bovine TB. As bovine TB is caused by bacteria, the correct treatment would be antibiotics.

The graph below shows trapped possums with TB (black) and without TB (grey). Graph (*a*) shows the control group and Graph (*b*) shows the vaccinated group.

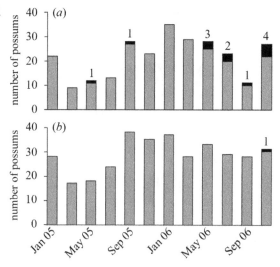

[Source: Tompkins D. M., Ramsey D. S. L., Cross M. L., Aldwell F. E., de Lisle G. W. and Buddle B. M. 2009. 'Oral vaccination reduces the incidence of tuberculosis in free-living brushtail possums.' Proc. R. Soc. B.2762987–2995 http://doi.org/10.1098/rspb.2009.0414.]

(c) Identify the month and year when the first case of TB was identified in the trial. **[1]**

....

(d) Calculate the percentage of possums with TB in November 2006 in the:
  (i) control group **[1]**

....

....

  (ii) vaccinated group. **[1]**

....

....

IDENTIFY

Here identify means that you need to give the correct answer from a range of possible answers (i.e. from Jan 05 to Sept 05).

ANSWER ANALYSIS

Use the key to identify that TB is the black boxes, then read down to see the month and year. Don't forget to put both the month **and** the year.

To read off the graph always draw a horizontal line so you can read accurately.

ANSWER ANALYSIS

This involves looking at the bottom graph. 1/31 × 100.

ANSWER ANALYSIS

This question requires you to look at the correct graph. The information above the graph states that the top graph is the control group. You then need to work out the last column on the right is November 2006 as the months are only labelled Jan, May, Sep, but the bars are Jan, March, May, July, Sep, November. Percentage with TB in the control group in November is 4 out of 27. The percentage = 4/27 × 100.

(e) Based on the graph, compare and contrast the infection rates with TB in the control and vaccinated groups. **[2]**

....

....

....

....

COMPARE AND CONTRAST

In a compare and contrast you need to give similarities and differences. Ensure you use the words 'both' and some comparatives such as 'higher' or 'more'.

ANSWER ANALYSIS

Look at the graphs for similarities and differences. You also need to be clear which group you are referring to when looking at differences. It helps to put the data from the graph axes in to help you with the answer, but you should not just state the data; rather you need to compare it, for example - higher/lower/earlier/later/more/less/twice/half, etc. You need to make two points for a 2-mark compare and contrast question. If looking for similarities and there are error bars that significantly overlap, there is no significant difference between the data. It helps to bullet point your answers. There is more than one possible correct answer for each mark.

A second long-term study investigated whether BCG-vaccinated calves would still be protected against *Mycobacterium bovis* two-and-a-half years after vaccination. The IFN-γ TB test responses measure the amount of interferon gamma antibody in response to the TB antigens.

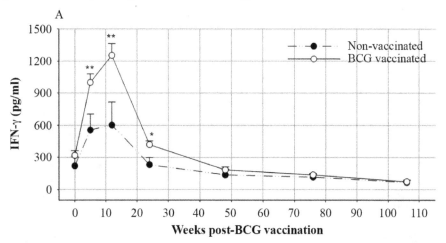

A

[Source: Parlane, N.A, Shu, D., Subharat, S., Wedlock, D.N., Rehm, B.H.A., (2014) 'Revaccination of Cattle with Bacille Calmette-Gue´rin Two Years after FirstVaccination when Immunity Has Waned, Boosted Protection against Challenge with Mycobacterium bovis'. PLoS ONE 9(9): e106519. doi:10.1371/journal.pone.0106519.]

(f) State the highest amount of IFN-γ in non-vaccinated calves. **[1]**

....

....

(g) Use the graph to outline the evidence that vaccination has the potential to prevent TB in cattle. **[2]**

....

....

....

....

(h) Using **all** the data, evaluate the use of vaccination to prevent the spread of TB to domestic livestock. **[3]**

....

....

....

....

....

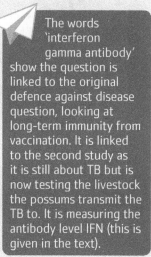

The words 'interferon gamma antibody' show the question is linked to the original defence against disease question, looking at long-term immunity from vaccination. It is linked to the second study as it is still about TB but is now testing the livestock the possums transmit the TB to. It is measuring the antibody level IFN (this is given in the text).

ANSWER ANALYSIS

A second graph is often given in Q1 which is a different type of graph from the first. It could also be a table or chart. In this case it's a line graph. The **state** question simply wants you to read off the answer from the correct line by using the key.

Don't forget to include the units, which are found by looking at the y axis.

ANSWER ANALYSIS

Here you need to look back at the topic in the introduction to the question for clues as well as using all the data in the different parts of the question. You will also need to use your knowledge about the topics covered such as vaccines and immunity to develop a conclusion. Be clear which graph or study the data has come from. Remember it links back to the text information at the start of the question and the syllabus link that you made.

ANSWER ANALYSIS

When evaluating data, you should look at the sample sizes and error bars. If the sample size is small or the study is over a short time, there may not be enough data for valid statistical analysis. If error bars are large there is a lot of variation in the data. Use knowledge or data or make a deduction from the data.

2. (a) Identify the ball and stick model shown below that was synthesized by Friedrich Wöhler and state the theory it falsified. **[1]**

Name: ....

Theory: ....

> Vitalism is the theory that living organisms have a vital force or a spark of life. This would suggest living carbon-containing molecules (organic molecules such as urea) could only be made by living organisms. This theory was falsified by Wöhler who synthesized organic urea from inorganic ammonium chloride and silver isocyanate.

(b) Define osmolarity and give its unit. **[2]**

....

....

....

....

(c) Distinguish between osmoconformers and osmoregulators. **[2]**

....

....

....

....

> Osmolarity is similar in terms to osmosis – you will need the words solute concentration, solution, litre and osmoles in your answer.

3. (a) Outline the structure of a nucleosome. **[2]**

....

....

....

(b) Nucleosomes help to regulate gene expression. Distinguish between methylation and acetylation in the control of gene expression. **[3]**

....

....

....

....

....

> In this outline question explain what chemicals a nucleosome consists of. How many histone proteins make up a nucleosome? How many types of histone proteins are involved? What does linker DNA do? A labelled diagram is useful here.

(c) An increase in the methylation of tumour suppressor genes can cause cancer. Suggest how. **[2]**

....

....

....

....

> Genes are transcribed more easily when the chromatin is not condensed. What is methylation and how does it affect transcription? What is acetylation and how does it affect transcription?

> Tumour suppressor genes need to be expressed to reduce mitosis.

(d) There are 25 cells in the image below. Calculate the mitotic index for the cells dividing. **[1]**

Count the number of cells in mitosis as a fraction of the total cells given in the question.

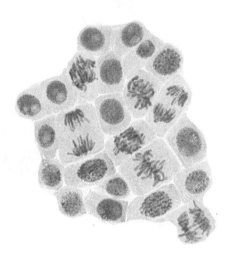

....

4. The image shows a cross-section through a coronavirus which is responsible for many respiratory issues such as colds, Flu and COVID-19. It consists of a protein coat and RNA.

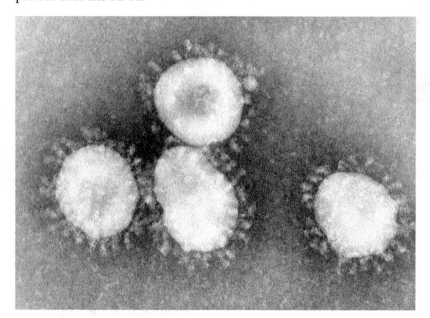

(a) State, with a reason, the type of microscope used to produce the image. **[1]**

....

....

Viruses are smaller than bacteria. Bacteria cannot be seen with a light microscope. The term cross-section should also give you a clue.

(b) Explain how vaccination can protect against coronavirus. **[5]**

....

....

....

....

....

....

....

You need to explain what a vaccine is made up of and how the vaccine affects our immune response antibodies.

....

(c) Outline why antibiotics would not be effective in treating people infected with coronaviruses. **[2]**

> You need to explain why antibiotics do not work on viruses.

....

....

....

5. The bluebell *Hyacinthoides non-scripta* is a woodland plant found in woodland in the United Kingdom. The diagram shows a flower from a hybrid bluebell produced from *Hyacinthoides non-scripta* and the Spanish bluebell, *Hyacinthoides hispanica*.

III. ....

II. ....

I. ....

(a) Label structures I, II and III. **[1]**

(b) Identify the phylum the bluebell belongs to. **[1]**

> Make sure you can label the male (anther and filament) and female (stigma, style, ovary) parts of a flower.

....

(c) Suggest, with a reason, how the plant is pollinated. **[2]**

> Brightly coloured flowers attract insects. How could an insect increase pollination between flowers?

....

....

....

....

(d) Outline the use of the binomial system of nomenclature in categorizing *Hyacinthoides non-scripta* and *Hyacinthoides hispanica*. **[2]**

> It is a system of classification introduced by Carl Linnaeus. What do the two (bi) name (nom) words represent?

....

....

....

....

Section B

Answer **two** questions from a choice of three. Up to one additional mark is available for the construction of your answers for each question.

6. Developments in the understanding of DNA have revolutionized medicine in the 21st century.

   (a) Draw a labelled diagram of a DNA nucleotide.   [3]

   Clearly draw a pentose sugar with a phosphate and base. Label with phosphate, deoxyribose sugar, nitrogenous base and a covalent bond.

   (b) Outline the function of enzymes in DNA replication.   [4]

   Think about the job of DNA gyrase, DNA polymerase III, DNA primase, DNA polymerase I and DNA ligase.

   ••• **OUTLINE**
   Give a brief account or a summary.

   (c) Describe the production of a genetically modified medicine and its use to treat a disease.   [8]

   What is genetic modification? How can enzymes be used to insert DNA into bacteria? Name a disease that can be treated with products made from genetically modified organisms.

7. Human beings contain approximately 5 litres of blood.

(a) State three substances transported in blood plasma. [3]

(b) Outline the coagulation cascade that occurs in blood. [5]

> The cascade is a metabolic pathway starting with clotting factors and ending with fibrin. You may find drawing a flow diagram useful in your answer.

(c) Explain the causes and consequences of hemophilia. [7]

> **EXPLAIN**
> Give a detailed account including the reasons or causes.

> State the type of inheritance. Identify the alleles involved. It is useful to draw a genetic cross between a carrier female and normal male, then identify the genotypes and phenotypes of the offspring.

8. Light energy from the sun is important to flowering plants.

(a) Draw a diagram of the structure of a chloroplast. **[3]**

Clearly label the double membrane, granum/thylakoid, stroma, starch grain, 70S ribosomes, loop of DNA and intermembrane space. You may wish to put on the approximate size.

(b) Explain the light-dependent stage of photosynthesis. **[7]**

State where it occurs, photolysis in Photosystem II, chemiosmosis, photophosphorylation, Photosystem I, reduction of NADP.

(c) Explain the control of flowering in angiospermophytes. **[5]**

Where are the meristems? How does flowering differ in short- and long-day plants?

Paper 3: Higher Level

- Set your timer for **1 hour and 15 minutes**
- The maximum mark for this examination paper is **[45 marks]**
- **Section A** – Answer ALL the questions
- **Section B** – Answer all of the questions in **ONE** of the options
- A calculator is required for this paper

Section A

1. The image below shows a scanning electron micrograph of a microsphere that is used to encapsulate the enzyme lactase.

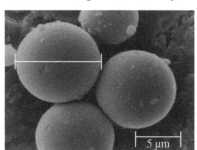

[Source: Zhao, W., Yang, R.J., Qian, T.T., Hua, X., Zhang, W.B., Katiyo, W. 'Preparation of Novel Poly(hydroxyethyl methacrylate-co-glycidyl methacrylate)-Grafted Core-Shell Magnetic Chitosan Microspheres and Immobilization of Lactase.' Int. J. Mol. Sci. 2013, 14, 12073-12089.]

(a) Calculate the magnification of the image. **[1]**

(b) Calculate the diameter of the microcapsule indicated. **[1]**

You can see by looking at the image that the diameter is more than the 5 µm scale bar but less than twice its size so your answer will be greater than 5 µm but less than 10 µm.

(c) State a use of lactase enzyme in industry. **[1]**

ANSWER ANALYSIS

To get 1 mark, you need to state **one** correct answer. There are a few different options you could state for the mark.

 Think about the food industry and how lactase enzyme might be used.

This is an example of a magnification question linked to immobilized enzymes.

Magnification = image size / actual size.

Actual size = 5 µm as it states the actual size of the scale bar underneath the picture.

To find the image size firstly measure the scale bar – labelled 5 µm – in mm = 16 mm.

As the actual size is in µm the image size also needs to be converted to µm.

Multiply by 1,000 to convert to µm so the scale bar image size matches the units on the actual size = 16,000 µm.

Now substitute into the formula.

! Don't forget to multiply the mm scale bar by 1,000 to convert to µm.

ANSWER ANALYSIS

There is another method you can use here. You can use ratios to find out how much bigger the diameter is than the scale bar.

Measure the diameter indicated in mm (27) and divide by the length of the scale bar (16) in mm then multiply your answer by 5 µm. (27/16 × 5 = 8.43).

If you can't work out the magnification you can get an 'error carried forward' mark as long as you write an answer to part (a) and divide by that answer for part (b).

2. A study investigated the effect of pH (**A**) and temperature (**B**) on free lactase and on immobilized lactase encapsulated in a microsphere.

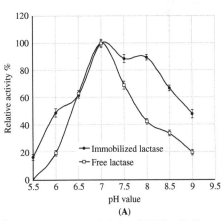 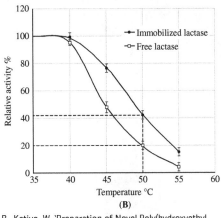

[Source: Zhao, W., Yang, R.-J., Qian, T.-T., Hua, X., Zhang, W.-B., Katiyo, W. 'Preparation of Novel Poly(hydroxyethyl methacrylate-co-glycidyl methacrylate)-Grafted Core-Shell Magnetic Chitosan Microspheres and Immobilization of Lactase.' Int. J. Mol. Sci. 2013, 14, 12073-12089.]

(a) Suggest **one** controlled variable for study **A**. [1]

(b) (i) Compare and contrast the effect of pH on immobilized lactase and free lactase. [2]

(ii) Suggest **one** advantage of immobilizing an enzyme such as lactase in industry. [1]

State similarities and differences using comparatives, e.g. 'both', 'higher', 'lower', 'longer'. Make sure you are looking at the pH graph (a).

3. Below is an image showing a dissection of a kidney.

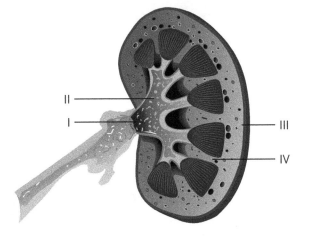

(a) Identify structures I, II, III and IV. [2]

I

II

III

IV

Learn the labels medulla, pelvis, cortex and ureter.

(b) State the name of the blood vessel that supplies the kidney with oxygenated blood. [1]

It is an artery that is connected with the kidney. The name for kidney is renal.

4. The image shows a section of a tobacco leaf chloroplast.

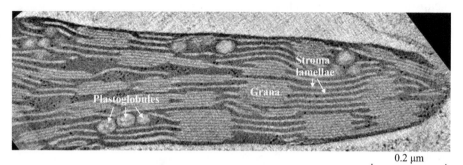

(Source: Yuv345, CC BY-SA 4.0)

(a) Calculate the magnification of the chloroplast. **[1]**

....

....

(b) State with a reason the type of microscope used to take this image. **[1]**

....

....

Melvin Calvin and colleagues used a lollipop experiment followed by autoradiography to identify the products of the light independent stages of photosynthesis. In the lollipop experiment, photosynthesising *Chlorella* algae were exposed to radioactive C^{14}. At different times following exposure to C^{14}, samples of the algae were dropped into a conical flask containing boiling ethanol.

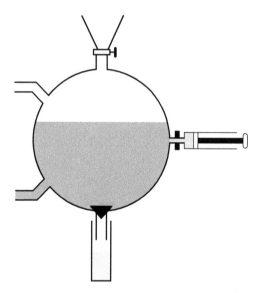

(c) Suggest why the algae was run into boiling ethanol. **[1]**

....

....

Autoradiograms showing the different carbon compounds in the alga *Chlorella* after exposure to $^{14}CO_2$ for 5 seconds (top image) and 30 seconds (bottom image). The darkness of the band indicates the amount of products.

Autoradiograms showing the labelling of carbon compounds in the alga *Chlorella* after exposure to $^{14}CO_2$

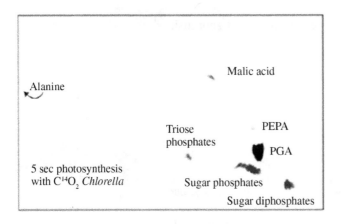

5 sec photosynthesis with $C^{14}O_2$ *Chlorella*

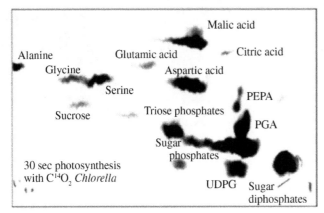

30 sec photosynthesis with $C^{14}O_2$ *Chlorella*

(Source: Bassham, T. A. (1965) *Photosynthesis: The path of carbon*. Plant Biochemistry, 2nd ed., J. Bonner and E. Varner, eds., Academic Press, New York, pp. 875–902.)

(d) Determine with a reason the first stable product of the light independent stage of photosynthesis. **[2]**

....

....

....

....

Section B

Option A – Neurobiology and behaviour

1. The image below shows an image series taken during neurulation of zebrafish embryos.

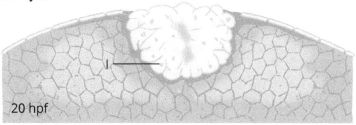

20 hpf

[Source: Adapted from Figure 1, Araya, C., Ward, L.C., Girdler, G.C. and Miranda, M. (2016), 'Coordinating cell and tissue behavior during zebrafish neural tube morphogenesis'. Dev. Dyn., 245: 197-208. https://doi.org/10.1002/dvdy.24304]

(a) Label I. **[1]**

....

(b) Describe the processes involved in neurulation. **[4]**

....

....

....

....

....

....

....

....

> Learn the labels of the different parts of an embryo such as ectoderm, mesoderm, endoderm, notochord, neural tube.

> Learn the stages involved from the infolding of the ectoderm, neural plate, neural crest and neural tube.

2. The image shows an fMRI of the brain that has been activated by alcohol.

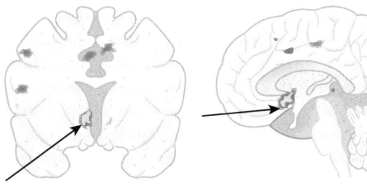

Back view   **Side view**

(a) Identify the area of the brain with the arrow. **[1]**

....

(b) Suggest which part of the brain is impaired causing loss of balance after consuming too much alcohol. **[1]**

....

> This part of the brain is involved in the pleasure/reward pathway.

> Which part of the brain affects balance?

(c) Explain the effect of alcohol on synapses. [4]

....

....

....

....

....

....

....

....

Think about how alcohol affects inhibitory GABA and excitatory glutamate neurotransmission. Why is alcohol addictive?

3. The diagram below shows the structure of the human eye.

You should learn the basic structure of the eye including pupil, iris, ciliary muscles, cornea, lens, fovea, retina and optic nerve.

State what brain death involves; explain how the pupil reflex works in terms of contraction and relaxing of circular and radial muscles. Outline what a reflex is. Explain how the pupil reflex can be used to assess the health of the brain stem and the implications this has.

(a) State the name of structures I, II, III, IV. [2]

I ....

II ....

III ....

IV ....

(b) Discuss the use of the pupil reflex to determine brain death. [4]

....

....

....

....

....

....

....

....

4. The blackcap (*Sylvia atricapilla*) is a common European migratory bird. It breeds in the summer and migrates to warmer areas for winter. Several migratory traits have been shown to be under genetic control in blackcaps, including timing of migration, direction, and duration/length of migration. A study compared genetic variation in different populations of blackcap. Traditional migratory routes involve migrating to south-western (SW) or south-eastern (SE) regions.

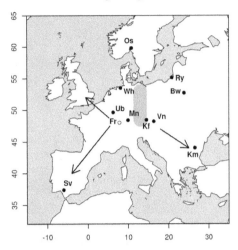

[Source: Mettler, R., Schaefer, H.M., Chernetsov, N., Fiedler, W., Hobson, K.A., Ilieva, M., et al. (2013) 'Contrasting Patterns of Genetic Differentiation among Blackcaps (Sylvia atricapilla) with Divergent Migratory Orientations in Europe.' PLoS ONE 8(11): e81365. https://doi.org/10.1371/journal.pone.0081365]

Key

Degrees East and North are shown on the X- AND Y-axis respectively. Letters on the graph represent the location of each population. The thick grey band shows the SW/SE divide. The dots represent the populations of blackcaps.

(a) State the type of behaviour shown by the blackcap. **[1]**

....

Is it innate or learned behaviour?

(b) State which direction the following populations migrated in. **[1]**

Bw: ....

Ub: ....

The study looked at the effect the longitudinal geographic distance had on the genetic distance between the populations.

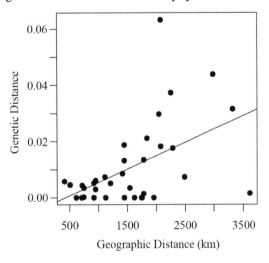

[Source: Mettler, R., Schaefer, H.M., Chernetsov, N., Fiedler, W., Hobson, K.A., Ilieva, M., et al. (2013) 'Contrasting Patterns of Genetic Differentiation among Blackcaps (Sylvia atricapilla) with Divergent Migratory Orientations in Europe.' PLoS ONE 8(11): e81365. https://doi.org/10.1371/journal.pone.0081365]

(c) Outline the trend between the geographic distance and genetic distance. **[1]**

....

....

In outline trend questions look at the pattern between increasing geographic distance and genetic distance.

(d) Use both graphs to suggest why the populations have not evolved into new species. **[1]**

....

....

Suggest why speciation may not occur. To answer you need to review the meaning of speciation which involves isolation of the populations and inhibition of interbreeding between populations.

A more recently established migratory route to the UK from Germany (Fr) shows that 11% of the Fr population fly in a northwest (NW) direction to the UK showing a change in migratory behaviour patterns. The pie charts show what percentage of migratory birds flew in which direction.

Key

Uebersyren, LU (Ub): NW = 1, SW = 42; Freiburg, DE (Fr): NW = 23, SW = 176, Unassigned (?) = 10; Kefermarkt, AT (Kf): South = 105; Białowieza, PL (Bw): South = 61.

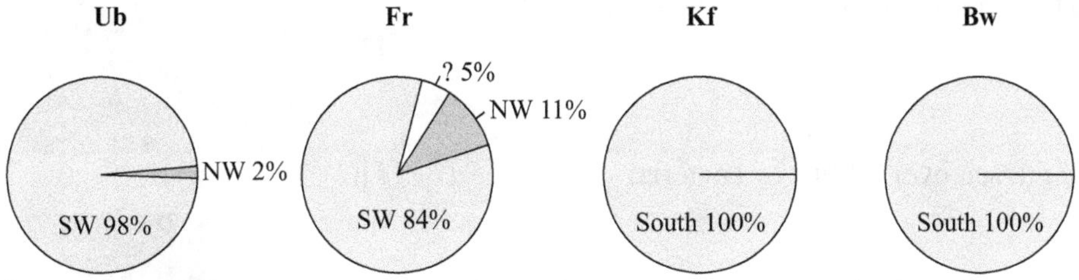

[Source: Mettler, R., Schaefer, H.M., Chernetsov, N., Fiedler, W., Hobson, K.A., Ilieva, M., et al. (2013) 'Contrasting Patterns of Genetic Differentiation among Blackcaps (Sylvia atricapilla) with Divergent Migratory Orientations in Europe'. PLoS ONE 8(11): e81365. https://doi.org/10.1371/journal.pone.0081365]

(e) Explain reasons why a population of the Fr population of birds have changed their migratory pattern. **[3]**

....

....

....

....

....

....

How can genetics and natural selection alter innate behaviour?

5. Evaluate methods used in brain research used to identify the function of different parts of the brain. **[6]**

....

....

....

....

....

....

....

....

....

....

....

....

....

....

You will need to compare the advantages and disadvantages of strokes, animal experiments and fMRI and PET scans leading to a conclusion as to which is best.

Option B – Biotechnology and bioinformatics

1. Antithrombin (AT) is a 432 amino acid protein that regulates thrombin and anticoagulant. Goats have been genetically modified to express antithrombin (ATryn) which is a glycoprotein in their milk.

(a) State the name given to the production of medicines by genetically modified animals. **[1]**

....

> It is a word that links biological and farming and pharmaceuticals.

(b) Suggest why goats were chosen over rabbits or bacteria. **[1]**

....

....

> Cows would also have been more suitable than rabbits!

(c) Outline the production of genetically modified antithrombin in goats. **[3]**

....

....

....

....

....

> Explain how the gene is isolated and expressed.

(d) Outline the use of microarrays in testing for mRNA in tissues. **[3]**

....

....

....

....

....

> What is a microarray? How is reverse transcriptase used? How are markers used and detected?

2. The image shows influenza virus H1N1 known as 'swine flu', a zoonotic virus transferred from pigs, causing flu-like symptoms in people it infected. Preliminary genetic research found that a membrane protein gene was similar to that of swine flu viruses present in US pigs since 1999, and matrix protein genes resembled versions present in European swine flu samples.

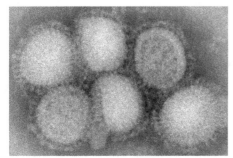

(a) Suggest how databases contributed to identification of the zoonotic origin as swine/pigs. **[2]**

Explain the use of powerful databases to compare DNA sequences.

(b) Explain how PCR is used in diagnosis of influenza. **[4]**

Explain the temperatures and cycles involved in the amplification of DNA.

3. A study investigated how effective two species of gram positive bacteria *Bacillus halotolerans* (1–1) and *Bacillus cereus* (T-04) were at removing crude oil from contaminated soil (5g/L crude oil) under different conditions of pH and salinity. Batch culture experiments were performed to investigate the percentage of crude oil removed from the soil by the bacteria (crude oil degradation rates) after 20 days.

[Source: Deng, Zhenshan & Jiang, Yingying & Chen, Kaikai & Gao, Fei & Liu, Xiaodong. (2020). 'Petroleum Depletion Property and Microbial Community Shift After Bioremediation Using Bacillus halotolerans T-04 and Bacillus cereus 1-1', *Frontiers in Microbiology*, 11. 10.3389/fmicb.2020.00353., www.researchgate.net/publication/339721752]

(a) State the name of the process that uses microorganisms to remove toxins from the environment. **[1]**

This process uses microorganisms to clean up a polluted environment.

(b) State what colour *Bacillus halotolerans* would be in a Gram stain. **[1]**

Look at the information in the question. Gram-positive bacteria are purple and Gram-negative are pink/red.

(c) Distinguish between batch and continuous fermentation. **[2]**

What is the difference between the words batch and continuous?

(d) Determine the maximum crude oil degradation for *Bacillus cereus* (T-04). **[1]**

Look at the graph.

(e) Outline the effect of soil pH on crude oil degradation for *Bacillus cereus* (T-04). **[1]**

....

....

....

....

> Analyse the shape of the graph for Zn.

(f) Evaluate the effectiveness of *Bacillus halotolerans* (1–1) and *Bacillus cereus* (T-04) in salty soils contaminated by crude oil. **[4]**

....

....

....

....

....

....

> Use all lines on the graph to compare if it is effective for each metal.

4. Discuss advantages and disadvantages of edible vaccines. **[6]**

....

....

....

....

....

....

....

....

....

....

....

....

> Make sure you include both advantages and disadvantages for a balanced argument.

Option C – Ecology and conservation

1. The climograph below shows the average monthly precipitation and temperature in Calcutta, India.

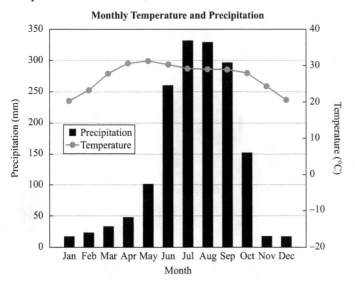

Monthly Temperature and Precipitation

(a) (i) Calculate the range of temperatures. [1]

....

Below is a climograph showing different biomes. Calcutta has an annual precipitation of 158 cm and an average temperature of 26 °C.

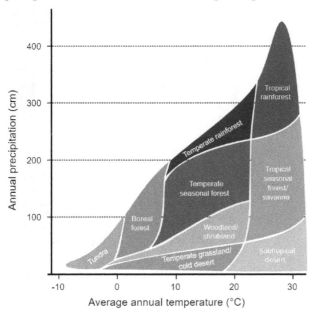

(ii) Define the term biome. [2]

....

....

....

....

(iii) Use the climograph to deduce Calcutta's biome. [1]

....

(b) Suggest a reason for the difference in net primary productivity between Taiga and Calcutta. [1]

....

....

2. The image shows a section of the Great Barrier Reef. The reef-building coral are formed of colonies of polyps glued together by calcium carbonate. Healthy coral is brightly coloured due to the association of the coral polyps with *Zooxanthellae*. Under certain stresses, the coral will bleach and turn white.

(a) State the trophic level of *Zooxanthellae*. [1]

....

> Trophic levels include producers, primary consumers, secondary consumers, etc.

(b) Describe the interaction between *Zooxanthellae* and coral polyps in reef-building corals. [2]

....

....

....

....

> How do they rely on each other for survival?

(c) Describe how climate change can affect coral reefs. [2]

....

....

....

....

> How do temperature and CO_2 concentration affect ocean chemistry? How does this affect the pH of the oceans? How do the coral polyps respond? What problem does this result in for the coral polyps?

(d) Outline how top-down limiting factors control the growth of algae on coral. [2]

....

....

....

....

> Explain how herbivores limit the population of autotrophs using the examples in the question.

3. The following data show the mean results of kick sampling in two rivers. Aquatic macro invertebrates are good indicators of biodiversity and habitat quality (Lee N. 2003). These insects inhabit aquatic ecosystems for most of their lives and are sensitive to different chemical and physical conditions (Cardoso et al, 2005; Anonymous, 2006). Their survival is closely linked to environmental conditions. An invertebrate with a pollution tolerance of 10 is highly sensitive to pollution, whereas an invertebrate with a tolerance of 1 is tolerant of pollution.

Species	Stonefly nymph	Mayfly nymph	Freshwater shrimp	Sludgeworm	Rat-tailed maggot	Water louse	Bloodworm
Sensitivity to pollution	Very sensitive	Sensitive	Tolerant	Tolerant	Tolerant	Very tolerant	Very tolerant
Image							

	Mean number of invertebrates sampled	
	River A	**River B**
Species		
Mayfly nymph	16	0
Stonefly nymph	15	0
Freshwater shrimp	5	3
Water louse	2	10
Bloodworm	0	12
Sludgeworm	0	15
Biotic index	8.5	

(a) State the meaning of a pollution indicator. **[1]**

....

....

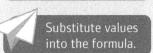

$$\text{Biotic index} = \frac{\Sigma\,(n \times a)}{N}$$

(b) State what N represents. **[1]**

....

(c) Calculate the biotic index of river B. **[1]**

....

(d) Determine which is the more polluted river. State your reason. **[2]**

....

....

....

....

4. Insectivorous plants, such as the Venus flytrap, trap and digest insects as a source of nitrogen in waterlogged soils.

(a) State the mode of nutrition of the Venus flytrap. **[1]**

....

What is an indicator species and how can it be used to determine pollution levels in the environment?

Remember that n = number of each species; N = total number of all species.

Substitute values into the formula.

Use the evidence in the table and your knowledge about invertebrates to determine the more polluted river.

Autotrophs make their own food by photosynthesis. Heterotrophs consume ALL food. A heterotroph is not the same as a plant that absorbs minerals!

(b) Outline the causes and consequences of excessive irrigation and poor
drainage on the nitrogen cycle. **[3]**

Draw the nitrogen cycle and use it to determine the effect of increased leaching on the cycle.

The graph below shows the change in worldwide phosphorus production from rocks.

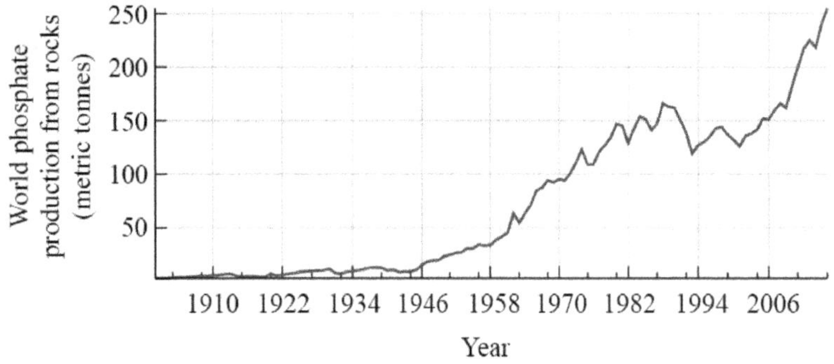

Think how phosphorous is needed in living molecules! Clue: DNA or RNA or ATP!

(c) State why organisms need phosphate. **[1]**

What environmental conditions release phosphates as pollution?

(d) Suggest a reason for the increase in phosphate production. **[1]**

Think about renewable or non-renewable resources.

(e) Outline a problem caused by increased phosphate demand. **[1]**

5. Explain the effects of resource availability on population growth curves. **[6]**

Describe the S-shaped sigmoid curve and its stages.

Option D – Human physiology

1. A study investigated the effect of adding a soluble dietary fibre called NVP to a
    breakfast cereal. The control group had no supplement of NVP, the experimental
    groups had a supplement of either 2.5 g or 5 g NVP. Blood glucose levels of
    the volunteers were measured every 15 minutes for 2 hours after the meal. The
    results are shown in the graph below. Error bars represent standard error.

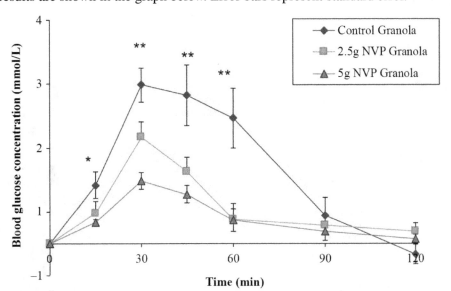

[Source: Jenkins, A.L., Kacinik, V., Lyon, M. et al. 'Effect of adding the novel fiber, PGX®, to commonly consumed
foods on glycemic response, glycemic index and GRIP: a simple and effective strategy for reducing post prandial
blood glucose levels - a randomized, controlled trial'. Nutr J 9, 58 (2010). https://doi.org/10.1186/1475-2891-9-58]

(a) Analyse the effect of NVP on blood glucose levels at 30 minutes
    after eating. [2]

> The blood glucose concentrations with a * and a ** are significantly different, while two blood glucose concentrations marked ** are not.

> The blood glucose levels need to be compared for each mass of NVP with the control. What do the error bars indicate?

(b) Suggest how the NVP causes its effect. [1]

> Remember that fibres cannot be digested by mammals.

(c) State two diseases that increasing fibre could help prevent. [2]

(d) State why fibre is important in the large intestine. [1]

(e) Outline the role of hormones on the appetite control centre in the
    hypothalamus. [2]

> The hormones leptin and insulin play a role in controlling the appetite.

2. The electron micrograph shows a section through a cell in the wall of an alveolus.

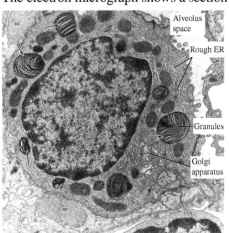

[Source: Junqueira, L.C. and Carneiro, J., *Basic Histology* 11th ed., McGraw-Hill, New York, 2005. p. 356.]

(a) Identify, with a reason, the type of cell shown. **[2]**

....

....

....

....

Remember, type I pneumocytes are flat cells responsible for gaseous exchange while type II pneumocytes are larger cells responsible for the production of surfactant.

(b) Suggest what the purpose of the granules are. **[2]**

....

....

....

....

3. The diagram below shows an electrocardiogram (ECG) of a heartbeat.

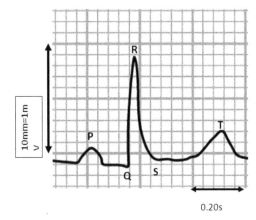

0.20s

(a) Outline the events in the cardiac cycle that are shown by the P, QRS and T waves. **[3]**

....

....

....

The cardiac cycle involves a series of atrial and ventricular systole and diastole.

(b) Calculate the time taken for the P to R interval. **[1]**

....

....

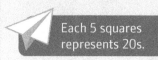

Each 5 squares represents 20s.

(c) Suggest why there is a gap between the P wave and the QRS wave. **[1]**

....

....

The image below shows the pressure changes in the aorta, atria and ventricle.

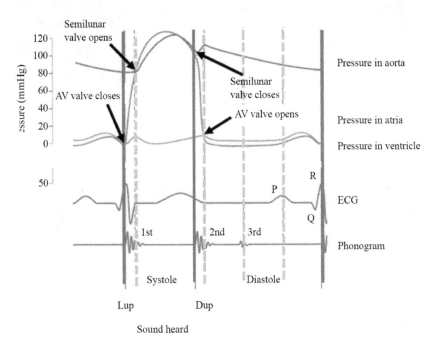

(d) A stethoscope can be used to listen to the heart sounds Lup-Dup. Use the image to deduce the cause of the sounds Lup and Dup in the diagram. **[2]**

> The sounds are indicated on the Phonogram shown on the image.

....

....

....

....

4. In 1904 Cristian Bohr investigated the effects of different partial pressures of carbon dioxide (5–80 mmHg) on the oxygen dissociation curve for hemoglobin. For each concentration of carbon dioxide he investigated how the partial pressure of oxygen in surrounding tissues affects the percentage saturation of hemoglobin with oxygen (oxy-hemoglobin). The graph below shows Bohr's results.

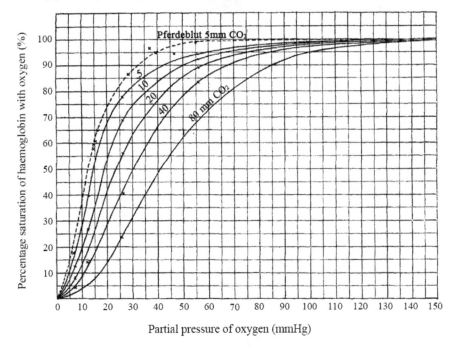

(a) State the partial pressure of oxygen for a CO_2 partial pressure of 40 mmHg that gives 50% saturation of hemoglobin with oxygen. **[1]**

....

(b) The alveoli have a partial pressure of oxygen of 104 mmHg and resting muscle tissue has a partial pressure of 40 mmHg. Outline the trend between increased CO_2 concentration and the percentage of oxy-hemoglobin in red blood cells in the lungs and the muscle tissues. **[2]**

At 104 mmHg, the difference in percentage saturation of oxygen is very small when comparing the values at different carbon dioxide concentrations. What about at 40 mmHg?

(c) Suggest, with a reason, what would happen to the oxygen dissociation curve during intense exercise. **[2]**

Does the curve move to the right or the left?

5. Outline the hormonal control of milk production. **[6]**

Set B

Are you ready to tackle Set B? There are fewer helpful tips and suggestions for this set so make sure you have done some revision before you try out these two papers.

Take at least a day's break between Paper 1 and Paper 2. Don't burn yourself out.

Have you remembered extra paper in case you run out of space?

Paper 1: Higher Level

- Set your timer for **1 hour**
- Each question is worth **[1]** mark
- The maximum mark for this examination paper is **[40 marks]**
- Answer ALL the questions

ANSWER ANALYSIS

You cannot have a calculator for Paper 1.

1. This image shows a giant algae *Acetabularia mediterranea*.

Why is it an exception to cell theory? [1]

- ☐ A. It does not come from pre-existing cells
- ☐ B. It only has one nucleus
- ☐ C. It is made up of 1 cell 100 mm in size
- ☐ D. It has hyphae

In this question you need to know the three principles of cell theory:

- All organisms are made of cells
- Cells are the smallest **unit** of life
- Cells come from pre-existing cells.

2. Sweet potato (*Ipomoea batatas*) and potato (*Solanum tuberosum*) cubes were placed in different strength salt solutions varying from 0.0 mol dm⁻³ to 1.0 mol dm⁻³ for 1 hour. The percentage change in mass of each tissue was calculated and plotted on the graph below.

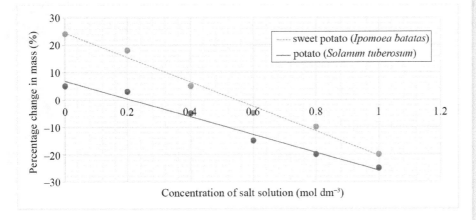

Osmosis is the movement of water from a hypotonic (dilute) solution to a hypertonic (concentrated) solution across a partially permeable membrane.

Use the graph to deduce which statement is correct. **[1]**

- ☐ A. At 0.4 mol dm^{-3} the potato is in a hypertonic solution, but the sweet potato is in a hypotonic solution
- ☐ B. At 0.2 mol dm^{-3} the solution is isotonic with the sweet potato
- ☐ C. The sweet potato has a lower concentration of solutes in its cytoplasm than potato
- ☐ D. Below 0.2 mol dm^{-3} both tissues are in hypertonic solutions

If water **enters** a cell, the cell will **gain** in mass. If water **leaves** a cell, the cell will **lose** mass. If the cell does not gain or lose mass, it has the **same concentration** as the solution (isotonic).

3. The diagram shows the concentration of cyclins DEAB during the cell cycle. What is the correct role of each cyclin? **[1]**

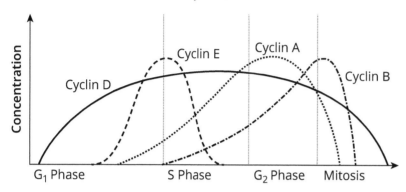

Make sure you know the stages of the cell cycle:
- **G1**: Cell grows and replicates the organelles
- **S**: Synthesizes new DNA by semi-conservative replication
- **G2**: Cell grows and prepares for division.

	Cyclin D	Cyclin E	Cyclin A	Cyclin B
☐ A.	Activates replication of DNA	Promotes mitotic spindle formation	Triggers cell to move to G1 and G2 phase	Prepares cell for replication
☐ B.	Prepares cell for replication	Activates replication of DNA	Promotes mitotic spindle formation	Triggers cell to move to G1 and G2 phase
☐ C.	Triggers cell to move to G1 and G2 phase	Prepares cell for replication	Activates replication of DNA	Promotes mitotic spindle formation
☐ D.	Promotes mitotic spindle formation	Triggers cell to move to G1 and G2 phase	Promotes mitotic spindle formation	Activates replication of DNA

There is a lot of transcription and translation in the growth stage of a cell as many proteins are made.

D: Triggers cells to pass into the G1 and G2 stage of the cell cycle

E: Prepares cell for S phase of replication

A: Activates replication of DNA (S phase)

B: Makes spindle fibres for prophase and prepares for mitosis.

4. The diagram shows the structure of linoleic acid.

H—C—C—C—C—C—C=C—C—C=C—C—C—C—C—C—C—C—C(=O)OH

What type of fatty acid is linoleic acid? **[1]**

- ☐ A. It is cis polyunsaturated
- ☐ B. It is monounsaturated
- ☐ C. It is saturated
- ☐ D. It is trans polyunsaturated

The hydrocarbon chain contains a carbon-to-carbon double bond (C=C). Both hydrogen atoms are on the same side of the C=C.

5. The diagram shows the digestion of amylose by amylase. **[1]**

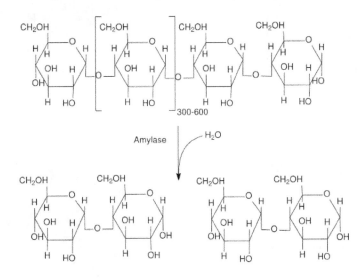

What type of reaction is occurring?

☐ A. Hydrolysis reaction breaking the glycosidic bond
☐ B. Condensation reaction breaking the glycosidic bond
☐ C. Condensation reaction breaking the ester bond
☐ D. Hydrolysis reaction breaking the ester bond

6. The image shows a bacterial protein. Identify structures X and Y. **[1]**

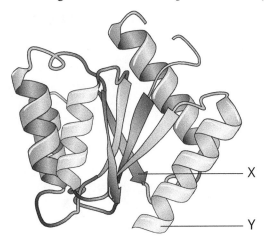

	X	Y
☐ A.	Alpha helix	Prosthetic group
☐ B.	Alpha helix	Beta pleated sheet
☐ C.	Beta pleated sheet	Alpha helix
☐ D.	Prosthetic group	Beta pleated sheet

7. Which of the following curves for an enzyme-catalysed reaction shows a non-competitive inhibitor? **[1]**

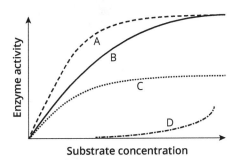

[Source: Adapted from: Berg, J.M., Tymoczko, J.L., Gatto, G.J.J., Stayer, L., *Biochemistry*. Eighth Edition. New York: W.H. Freeman and Company, 2015]

Starch is being split into maltose. Water is being added to break the glycosidic bond.

Condensation involves the removal of water in anabolic reactions, whereas hydrolysis involves adding water in catabolic reactions.

Ester bonds are found in fats. Peptide bonds are in polypeptides and glycosidic bonds are in sugars.

The secondary structures are made in different ways:
- **Beta pleated sheet**: The primary structure is folded into a zig-zag shape
- **Alpha helix**: The primary structure is coiled into a helical shape.

C is non-competitive as the inhibitor does not compete for the active site. Instead, the inhibitor binds to the allosteric site so the active site changes shape and cannot bind to the enzyme no matter how much substrate is added.

A shows competitive inhibition as the inhibitor is complementary in shape to the active site, so binds to the active site if the competitive inhibitor collides with the enzyme. Increasing the substrate concentration reduces the chance of the competitive inhibitor colliding with the active site and increases the chance of the substrate colliding with the active site, 'out competing' the inhibitor.

8. Identify the molecule below. **[1]**

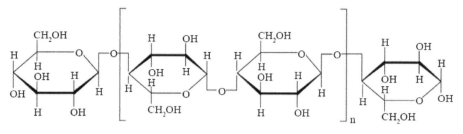

☐ A. Glycogen

☐ B. Amylopectin

☐ C. Amylose

☐ D. Cellulose

Cellulose is a polysaccharide made up of alternatively oriented beta glucose. It is unbranched.

Amylose and amylopectin are both forms of starch. They are made up of alpha glucose. Amylose is unbranched and only contains 1–4 glycosidic bonds whereas amylopectin is branched and contains 1–4 and 1–6 glycosidic bonds.

9. Hershey and Chase experimented to determine the chemical nature of genetic material. They cultured viruses called bacteriophages. These were labelled with either radioactive isotopes of sulphur ^{32}S (which attached to the virus's protein coats) or radioactive phosphorus ^{12}P (which attached to the DNA sugar–phosphate backbone). Hershey and Chase infected bacterial cells with the virus and following centrifugation they measured the radioactivity in both the cell and the fluid around the cell (supernatant). Their experiment is summarized below. **[1]**

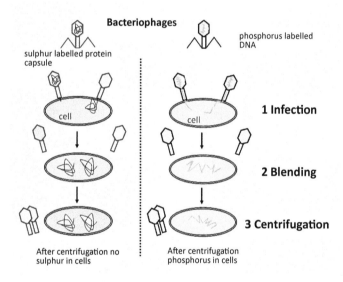

Glycogen is how glucose is stored in muscles and the liver. It is similar in structure to amylose but has more branches. It contains 1–4 glycosidic bonds only.

Bacteriophages are viruses that invade bacteria. The viruses pass their genetic material into the host bacteria and make lots of viruses inside the bacterial cell.

Hershey and Chase chose to use S and P as sulphur is found in amino acids and not DNA, while phosphorus is found in DNA and not proteins.

Which statement below correctly interprets the evidence supporting DNA as the genetic material?

☐ A. The bacterial pellet contained radioactive phosphorus so contains DNA

☐ B. The bacterial pellet contained radioactive sulphur so contains DNA

☐ C. The supernatant contained radioactive sulphur so contains DNA

☐ D. The supernatant contained radioactive phosphorus so contains DNA

10. Which of the following can increase transcription, allowing genes to be expressed? **[1]**

	Cytosines	Histones	Chromatin
☐ A.	Unmethylated	Acetylated	Not condensed
☐ B.	Unmethylated	Methylated	Not condensed
☐ C.	Methylated	Acetylated	Condensed
☐ D.	Methylated	Methylated	Condensed

- **Methylation** reduces gene expression
- **Acetylation** increases gene expression
- **Post-transcriptional modification** removes introns altering the mRNA before translation into proteins.

11. What is the correct definition for translation and where does it occur? Choose the correct row. **[1]**

		Location	Definition
☐	A.	Nucleus	DNA is used as a template to make mRNA
☐	B.	Nucleus	mRNA is used as a template to make proteins
☐	C.	Cytoplasm	DNA is used as a template to make mRNA
☐	D.	Cytoplasm	mRNA is used as a template to make proteins

> **Transcription** is the copying of the DNA code into mRNA.

> **Translation** is the decoding of mRNA into a sequence of amino acids.

12. Below is a section of DNA in the beta globin gene Hb^A that has mutated to form a different allele Hb^S.

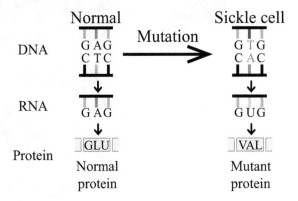

Which statement is the correct explanation and consequence of the mutation GAG → GTG? **[1]**

- ☐ A. Non-disjunction of chromosome 21 in meiosis causing Down syndrome
- ☐ B. Base substitution of A to T resulting in amino acid valine causing sickle cell anemia
- ☐ C. Base substitution of T to A resulting in amino acid glutamic acid causing sickle cell anemia
- ☐ D. Non-disjunction of chromosome 18 in meiosis causing Down syndrome

> Hemoglobin is found in red blood cells and transports oxygen. It is a conjugated protein consisting of four polypeptides (two alpha globin and two beta globin), each attached to iron (heme) prosthetic groups.

> A base substitution mutation (A to T) on the 6th codon of the sense strand for beta globin results in the normal codon GAG being replaced with GTG. During translation GAG codes for the more polar glutamic acid amino acid and GUG codes for the less polar valine. This means the normal mRNA GAG is replaced with GUG in the primary structure of the polypeptide. As a result, the beta hemoglobin becomes sickle shaped in low oxygen concentrations, resulting in sickle cell anemia, chronic fatigue and severe pain.

> Beta globin codes for 147 amino acids.

> The two alleles are co-dominant: Hb^A is the normal allele and Hb^S is the sickle cell allele. If a person has sickle cell disease, they have the genotype Hb^{S}Hb^S. If they are a carrier of sickle cell disease, they have the genotype Hb^{A}Hb^S and have resistance to malaria. Hb^{A}Hb^A is the normal phenotype.

13. What name is given to a non-coding section of DNA with a function? **[1]**

- ☐ A. Exon
- ☐ B. Intron
- ☐ C. Promoter
- ☐ D. Codon

> • An **exon** is part of a gene that is both transcribed and translated.
> • An **intron** is a nucleotide sequence within a gene that is transcribed but not translated as they are removed by splicing.
> • A **promotor** is a non-coding part of DNA found upstream of the gene that initiates transcription of a gene.
> • A **codon** is the name given to three nucleotides that code for a specific amino acid.

14. Below is a Punnett grid showing the inheritance of skin colour in humans. The parents are both heterozygous (AaBbCc × AaBbCc).

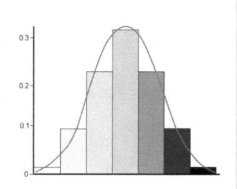

	ABC	ABc	AbC	aBC	Abc	aBc	abC	abc
ABC	6	5	5	5	4	4	4	3
ABc	5	4	4	4	3	3	3	2
AbC	5	4	4	4	3	3	3	2
aBC	5	4	4	4	3	3	3	2
Abc	4	3	3	3	2	2	2	1
aBc	4	3	3	3	2	2	2	1
abC	4	3	3	3	2	2	2	1
abc	3	2	2	2	1	1	1	0

What could explain these results? [1]

☐ A. Multiple alleles

☐ B. Autosomal dominant trait

☐ C. Sex-linked recessive trait

☐ D. Polygenic inheritance

The 1, 6, 15, 20, 15, 6, 1 phenotypic ratio links to Pascal's triangle in maths.

Don't confuse polygenic inheritance with multiple alleles. Multiple alleles refers to one gene at one locus on one chromosome that has more than two different alleles (e.g. blood type has three alleles: IA, IB and i). The phenotypes show discontinuous variation as people are either phenotype A, B, AB or O.

In polygenic inheritance a single characteristic is controlled by two or more genes. The genes are different and are found at different loci, often on different chromosomes. The alleles are co-dominant so there is an additive effect. In this example, there are three genes (A, B and C) for making the protein melanin.

With two heterozygous parents (AaBbCc × AaBbCc) there is the following phenotypic ratio: 1, 6, 15, 20, 15, 6, 1. The large number of phenotypes mean the traits start to 'blend' into each other as a result of environmental differences.

15. In fruit flies (*Drosophila melanogaster*) the allele for wild-type body colour is dominant to black body, and the allele for normal wings is dominant to vestigial wings.

Flies homozygous for wild-type body colour and normal wings were crossed with flies with black body and vestigial wings. 1,000 of the offspring survived to maturity and the number of offspring with each phenotype were recorded in the table below.

Phenotype of offspring	Frequency of offspring
Wild-type body colour, normal wings	460
Wild-type body colour, vestigial wings	50
Black colour body, normal wings	50
Black colour body, vestigial wings	440

What statistical test could be done to determine if the observed results differ from the expected 1:1:1:1 ratio? [1]

☐ A. T-test

☐ B. Chi-squared test

☐ C. Correlation

☐ D. Standard deviation

- A **T-test** looks for significant differences between the means of two populations (e.g. the mean height of population A is significantly higher than the mean height of population B).
- A **chi-squared** test compares counts of observed values to the expected values.
- A **correlation** looks at relationships between two variables (e.g. as one increases the other decreases).
- Standard deviation gives an idea how much a set of repeats in data are spread about the mean value of the data. It is often used as error bars on graphs.

A small standard deviation means the data are close to the mean. The rule is that 68% of data lie within one standard deviation of the mean.

16. Mendel investigated dihybrid crosses in peas. The allele for round seed shape R is dominant to wrinkled r. The allele for yellow seed colour Y is dominant to green y. Two heterozygotes were crossed. What is the expected phenotypic ratio in the offspring? **[1]**

- ☐ A. 1 yellow round : 1 yellow wrinkled : 1 green round : 1 green wrinkled
- ☐ B. 9 yellow round : 3 yellow wrinkled : 3 green round : 1 green wrinkled
- ☐ C. 3 yellow round : 1 yellow wrinkled
- ☐ D. 3 green round : 1 green wrinkled

> In a dihybrid cross, if a heterozygote is crossed with another heterozygote the ratio will be 9:3:3:1 (with the 9 being the double dominant and the 1 being the double recessive).

17. In the fruit fly *Drosophila melanogaster* the genes for body colour and wing type are linked. The allele for ebony body colour, (e), is recessive to the allele for yellow wild-type body colour (E). The allele for vestigial (vg) wings is recessive to the allele for longer wild-type wing (Vg).

What are all the possible gametes that an individual with the genotype below could form?

$$\frac{\overline{E \quad Vg}}{e \quad vg}$$

[1]

> Crossing over occurs in prophase I meaning that there will be a small number of the recombinants.

- ☐ A. $\overline{E \quad Vg}$ $\overline{e \quad vg}$
- ☐ B. $\overline{E \quad Vg}$ $\overline{E \quad vg}$ $\overline{e \quad Vg}$ $\overline{e \quad vg}$
- ☐ C. $\overline{E \quad e}$ $\overline{Vg \quad vg}$
- ☐ D. $\overline{E \quad e}$ $\overline{Vg \quad vg}$ $\overline{E \quad Vg}$ $\overline{e \quad vg}$

18. A food chain is shown below. Which creatures are the primary consumer? **[1]**

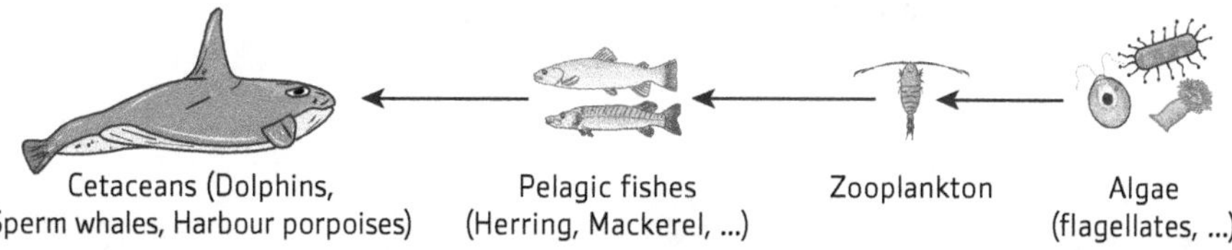

Cetaceans (Dolphins, Sperm whales, Harbour porpoises) ← Pelagic fishes (Herring, Mackerel, ...) ← Zooplankton ← Algae (flagellates, ...)

- ☐ A. Algae
- ☐ B. Zooplankton
- ☐ C. Pelagic fishes
- ☐ D. Cetaceans

> A producer will always be the first trophic level. It is an organism that can produce food from inorganic compounds and is also called an autotroph. A primary consumer will always receive its organic compounds by feeding off of autotrophs.

19. Which organisms produce methane in anaerobic environments such as rice fields? **[1]**

- ☐ A. Eubacteria
- ☐ B. Archaea
- ☐ C. Eukaryotes
- ☐ D. Prokaryotes

> **Archaea** contains extremophiles (such as methanogens that make the gas methane).

> **Eukaryota** contains eukaryotes (such as plants, animals, protoctista and fungi).

20. What is the correct location and process at positions I, II, III? **[1]**

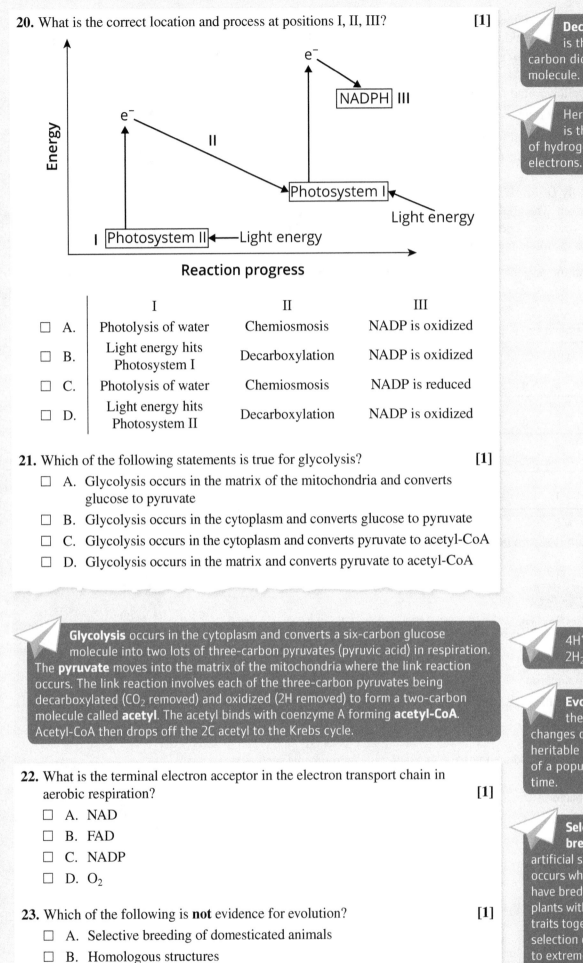

Decarboxylation is the removal of carbon dioxide from a molecule.

Here **reduction** is the addition of hydrogen and of electrons.

		I	II	III
☐	A.	Photolysis of water	Chemiosmosis	NADP is oxidized
☐	B.	Light energy hits Photosystem I	Decarboxylation	NADP is oxidized
☐	C.	Photolysis of water	Chemiosmosis	NADP is reduced
☐	D.	Light energy hits Photosystem II	Decarboxylation	NADP is oxidized

21. Which of the following statements is true for glycolysis? **[1]**

☐ A. Glycolysis occurs in the matrix of the mitochondria and converts glucose to pyruvate

☐ B. Glycolysis occurs in the cytoplasm and converts glucose to pyruvate

☐ C. Glycolysis occurs in the cytoplasm and converts pyruvate to acetyl-CoA

☐ D. Glycolysis occurs in the matrix and converts pyruvate to acetyl-CoA

Glycolysis occurs in the cytoplasm and converts a six-carbon glucose molecule into two lots of three-carbon pyruvates (pyruvic acid) in respiration. The **pyruvate** moves into the matrix of the mitochondria where the link reaction occurs. The link reaction involves each of the three-carbon pyruvates being decarboxylated (CO_2 removed) and oxidized (2H removed) to form a two-carbon molecule called **acetyl**. The acetyl binds with coenzyme A forming **acetyl-CoA**. Acetyl-CoA then drops off the 2C acetyl to the Krebs cycle.

$4H^+ + 4e^- + O_2 \rightarrow 2H_2O$

Evolution is the cumulative changes of the heritable characteristics of a population over time.

22. What is the terminal electron acceptor in the electron transport chain in aerobic respiration? **[1]**

☐ A. NAD

☐ B. FAD

☐ C. NADP

☐ D. O_2

23. Which of the following is **not** evidence for evolution? **[1]**

☐ A. Selective breeding of domesticated animals

☐ B. Homologous structures

☐ C. Industrial melanism

☐ D. Differences in chromosome numbers between species

Selective breeding or artificial selection occurs when humans have bred animals or plants with desirable traits together. The selection of traits has led to extreme differences in phenotypes such as the huge variety in dog breeds and the large differences between dog breeds and their wild ancestor the wolf.

24. The image shows the phylogenetic tree for insects that feed on plants. Which insects are most closely related to Aphididae? **[1]**

- ☐ A. Sternorrhyncha
- ☐ B. Adelgidae
- ☐ C. Psylloidea
- ☐ D. Coccoidea

25. Which feature below belongs to the phylum Filicinophyta? **[1]**
- ☐ A. No xylem or phloem and reproduces from spores
- ☐ B. Has xylem and phloem and reproduces from spores
- ☐ C. Has xylem and phloem and reproduces from seeds in cones
- ☐ D. Has xylem and phloem and reproduces from seeds in flowers

26. The large ground finch (*Geospiza magnirostris*) over time has undergone natural selection to develop a wider and shorter beak adapted to crack nuts. What type of natural selection does this show? **[1]**
- ☐ A. Directional natural selection
- ☐ B. Stabilizing natural selection
- ☐ C. Disruptive natural selection
- ☐ D. Diverging natural selection

27. Which hormones in the menstrual cycle are part of a negative feedback mechanism? **[1]**
- ☐ A. LH stimulates progesterone
- ☐ B. Oestrogen inhibits FSH
- ☐ C. LH inhibits progesterone
- ☐ D. Oestrogen stimulates LH

28. Which statement about meristems is correct? **[1]**
- ☐ A. Auxin decreases mitosis and cell elongation at the tips of shoots and lateral buds
- ☐ B. Auxin increases mitosis and cell elongation at the tips of shoots and lateral buds
- ☐ C. Auxin increases mitosis and cell elongation at the tips of shoots but inhibits lateral buds
- ☐ D. Auxin increases mitosis and cell elongation at the tips of lateral buds but inhibits shoots

A closely related species shares a recent node with a close relative. More distantly related species are linked by a series of nodes.

- **Bryophyta** have no vascular tissue but do reproduce by spores.
- **Filicinophyta** have vascular tissue and also reproduce by spores under the leaf.
- **Coniferophyta** and **Angiospermophyta** both have vascular tissues and seeds (not spores).

There are three types of natural selection:
- **Directional** selection occurs if an extreme phenotype is favoured
- **Stabilizing** selection occurs if the intermediate phenotype is favoured
- **Disruptive** selection is where the extreme phenotypes are favoured over the intermediate.

Meristems are plant growth regions. Apical meristems are found at the tips of roots, shoots (terminal bud) and lateral (axillary) buds. Auxins made in the terminal bud in the shoot promote the shoot apex to undergo cell division and elongation. This allows the plant to grow upwards towards the light.

Auxin inhibits the lateral (axillary) buds, preventing branching. As the shoot apex grows, the distance increases between the terminal bud and the axillary bud so the inhibition decreases. This allows the lateral bud meristems to form branches.

29. Which statement is true about the antibiotic streptomycin? **[1]**

- ☐ A. It can be used to treat the influenza virus
- ☐ B. It damages eukaryotic cells without damaging the host cells
- ☐ C. It will not work on viruses because viruses do not have their own metabolism
- ☐ D. It kills all types of bacteria

Antibiotics work by attacking part of bacterial cells that are not present in eukaryotic cells, such as the peptidoglycan cell wall. This means they kill the prokaryotes without harming the eukaryotes.

Viruses do not have their own metabolism so they invade hosts cells. As a result, viruses can't be destroyed without damaging the host cell. This means antibiotics will not work on viruses.

30. Which factor reduces the rate of transpiration? **[1]**

- ☐ A. High temperature
- ☐ B. High humidity
- ☐ C. High wind
- ☐ D. High light level

Transpiration increases if the water vapour diffuses more quickly from the leaf. A high temperature increases the kinetic energy of the water vapour molecules so diffusion will be faster. Wind will reduce the water molecules on the outside of the leaf so the concentration gradient between the inside and outside will be steeper and diffusion faster. A high light level will mean stomata are open so transpiration increases. High humidity will increase the water vapour outside the stomata, which will make the concentration gradient shallower so diffusion will be slower.

31. Below is a molecule showing the secondary structure of an RNA molecule.

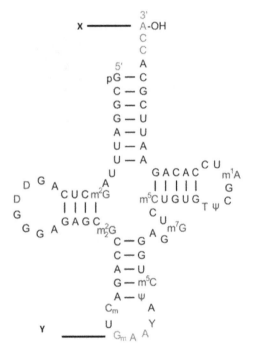

Which row correctly names the molecule and describes the functions of the parts labelled X and Y? **[1]**

mRNA is a single molecule of RNA made by transcription. mRNA has no sections of complementary base pairing.

tRNA is folded into a cloverleaf secondary structure. The bottom loop consists of three bases called the anticodon that complementary base pair with mRNA. The 3-carbon end consists of a CCA tail that attaches to a specific amino acid. Then the tRNA activating enzyme attaches the amino acid in place to allow translation.

	Name	Function of X	Function of Y
☐ A.	mRNA	Binds to small ribosome subunit	Binds to codon on tRNA
☐ B.	tRNA	Binds to amino acid	Binds to codon on mRNA
☐ C.	mRNA	Binds to DNA	Binds to anticodon on tRNA
☐ D.	tRNA	Binds to helicase	Binds to anticodon on mRNA

32. Below is a section of mRNA containing introns and exons. **[1]**

| Exon 1 | Intron 1 | Exon 2 | Intron 2 | Exon 3 | Intron 3 | Exon 4 | Intron 4 |

Which could be a mature mRNA molecule made after splicing?

☐ A.

| Exon 1 | Intron 1 | Exon 2 | Intron 2 |

☐ B.

| Intron 1 | Intron 2 | Intron 3 | Intron 4 |

☐ C.

| Exon 1 | Exon 2 | Exon 3 | Exon 4 |

☐ D.

| Exon 1 | Intron 1 | Exon 3 | Intron 3 |

Splicing involves removing introns from immature mRNA to make mature mRNA. Only the exons are expressed after translation. Different tissues may express different exons resulting in one gene being able to make more than one polypeptide.

33. In the production of urine, what happens if the solute concentration of the blood is too high (hypertonic)? **[1]**

☐ A. The collecting duct becomes less permeable to water

☐ B. The pituitary gland secretes less ADH

☐ C. Aquaporins form on the membrane of the collecting duct

☐ D. A large volume of dilute urine is formed

If the blood is hypertonic then osmoreceptors in the hypothalamus shrink, stimulating the release of ADH from the posterior pituitary. ADH travels in the blood to the collecting duct of the kidney. It stimulates more aquaporins, which makes it more permeable to water. This means more water is reabsorbed from the collecting duct by osmosis so a small volume of concentrated urine is made.

34. How do insects carry out osmoregulation and what is their excretory product? **[1]**

	Osmoregulation	Excretory product
☐ A.	Nephrons	Amino acids
☐ B.	Malpighian tubules	Urea
☐ C.	Nephrons	Ammonia
☐ D.	Malpighian tubules	Uric acid

Malpighian tubules convert ammonia from the haemolymph into uric acid. Water enters the Malpighian tubules by osmosis and flushes the uric acid to the hindgut. The water is reabsorbed, and the uric acid excreted as a paste.

35. The diagram shows an electron micrograph of striated muscle. What is the structure labelled X? **[1]**

☐ A. Sarcomere
☐ B. Myofibril
☐ C. Muscle fibre
☐ D. Sarcolemma

Sarcomeres join end to end to form a muscle cell called a myofibril with a diameter of 1 μm. On myofibrils it is easy to identify the different banding patterns. A muscle fibre is made up of many myofibrils with many nuclei and has a larger diameter of 100 μm.

36. What are the names of X, Y and Z in the image below? **[1]**

		X	Y	Z
☐	A.	Triceps	Ulna	Ligament
☐	B.	Biceps	Radius	Tendon
☐	C.	Triceps	Radius	Ligament
☐	D.	Biceps	Ulna	Tendon

37. Which row contains the correct statements about arteries, veins and capillaries? **[1]**

		Artery	**Capillary**	**Vein**
☐	A.	Thick muscular wall with elastic fibres and muscle	Wall one cell thick	Valves present
☐	B.	Always oxygenated	Gas exchange occurs here	Always deoxygenated
☐	C.	Valves present	Always oxygenated	Thick muscular wall with elastic fibres and muscle
☐	D.	Always deoxygenated	Valves present	Wall one cell thick

> Arteries are usually oxygenated but there are exceptions. For example, the pulmonary artery carries deoxygenated blood from the heart to the lungs and the umbilical artery carries deoxygenated blood from the fetus through the umbilical cord.

38. Which statement is true during inhalation? **[1]**

☐ A. Both the external intercostal muscles and diaphragm contract

☐ B. Both the internal intercostal muscles and diaphragm contract

☐ C. The external intercostal muscles contract and the diaphragm relaxes

☐ D. The internal intercostal muscles contract and the diaphragm relaxes

> During inhalation:
> • The **external intercostal** muscles contract and pull the ribs up and out.
> • The **diaphragm** contracts and flattens.
> • The volume inside the **thorax** increases, which reduces the pressure.
> • Air flows into the lungs from the higher atmospheric pressure to the lower pressure inside the lungs.

Ligaments connect bone to bone.

Tendons connect muscle to bone.

The ulna is underneath the radius.

Veins go **in**to the heart. **A**rteries go **a**way from the heart.

Valves are present in veins to stop the slow-moving low-pressure blood from flowing backwards.

Capillary walls are one cell thick to allow a short diffusion distance for fast diffusion.

39. What type of cell is made when plasma cells are fused with tumour cells
in the manufacture of monoclonal antibodies? **[1]**

☐ A. Sarcoma cells

☐ B. Hybridoma cells

☐ C. Memory cells

☐ D. Red blood cells

> Monoclonal antibodies are clones of the same parent cell and make a specific antibody. To make a monoclonal antibody, an animal is injected with an antigen and it makes B lymphocytes that are specific to this antigen. A small piece of the spleen is removed from the animal.

> Tumour cells cannot make antibodies but can divide by mitosis endlessly. A hybridoma cell occurs when the plasma cell is fused with the tumour cell producing a cell that has the benefit of both endless cell division and the ability to make antibodies. Therapeutic uses of monoclonal antibodies include treating rabies and cancer. Diagnostic uses include the detection of malaria, HIV and cancers.

40. This image is of a light micrograph of a seminiferous tubule of testes.

> Spermatogenesis involves the production of sperm (spermatozoa) from the initial germinal epithelium at the edge of the tubule to the spermatozoa at the centre of the tubule.

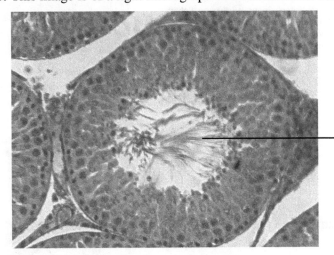

—X

What is the name of the cells labelled X? **[1]**

☐ A. Spermatogonia

☐ B. Spermatocytes

☐ C. Spermatids

☐ D. Spermatozoa

> The germinal epithelium on the outside of the seminiferous tubule divides by mitosis, making diploid cells that grow into primary spermatocytes. Theses diploid primary spermatocytes undergo the first meiotic division to make two haploid secondary spermatocytes. The haploid secondary spermatocytes undergo a second meiotic division to form two haploid spermatids. Sertoli (nurse) cells associate with the spermatids to enable them to differentiate into haploid sperm cells called spermatozoa.

Paper 2: Higher Level

- Set your timer for **2 hours and 15 minutes**
- The maximum mark for this examination paper is **[72 marks]**
- **Section A** – answer ALL the questions
- **Section B** – answer two questions
- A calculator is needed for this paper

Section A

1. Huntington's disease is a genetic disorder that leads to the degeneration of neurons. Symptoms include involuntary and prolonged muscle contraction.

   A study investigated skeletal muscle fibres from mice with Huntington's disease (HD) and a control group (WT). The minimum current (nA) required to depolarize the fibre and generate an action potential in the muscle fibres was recorded (minimum AP stimulus).

   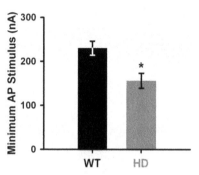

   [Source: Christopher W. Waters, Grigor Varuzhanyan, Robert J. Talmadge, and Andrew A. Voss, 'Huntington disease skeletal muscle is hyperexcitable owing to chloride and potassium channel dysfunction' (Proceedings of the National Academy of Sciences May 2013, 110 (22) 9160-9165; DOI: 10.1073/pnas.1220068110)]

   (a) State the minimum current needed to produce an action potential in: **[1]**

      (i) control mice

   ....

      (ii) HD mice.

   ....

   > Minimum 0.5-ms depolarizing stimulus current (mean ± SEM) needed to trigger an action potential (AP) in WT (n = 17) and HD (n = 12) fibres.

   > Use the information in the introduction to the question to make sure you are reading off from the correct line. Always use a ruler to draw across.

   (b) Calculate the percentage change in the minimum action potential stimulus. **[1]**

   ....

   > Percentage change = (difference / original) × 100

   The action potential of the mice's muscle fibres was recorded by stimulating the muscle fibre at 0 s and recording the membrane potential in millivolts (mV).

   The graph shows the membrane potential (mV) for the muscle fibre at resting, depolarization and repolarization after a stimulus at 0 s.

   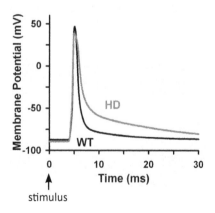

   [Source: Christopher W. Waters, Grigor Varuzhanyan, Robert J. Talmadge, and Andrew A. Voss, 'Huntington disease skeletal muscle is hyperexcitable owing to chloride and potassium channel dysfunction' (Proceedings of the National Academy of Sciences May 2013, 110 (22) 9160-9165; DOI: 10.1073/pnas.1220068110)]

   (c) State the resting membrane potential for both neurons. **[1]**

   ....

   ....

   > Each graduation = 25 V – this is half-way down the 25 V line.

(d) Compare and contrast the changes in action potentials for mice with and without Huntington's disease. **[2]**

....

....

....

....

Link to the terms depolarization and repolarization. Look at the shape and values – look for similarities and differences. Write what you see. Use comparatives such as both, higher, longer, shorter.

(e) Deduce whether Huntington's affects the depolarization **or** repolarization of the muscle fibre. **[2]**

....

....

....

....

You should be able to identify the resting potential, depolarization, and repolarization of a neuron.

The study investigated the peak current ($_{ICl}$ μA/cm^2) flowing through chloride and potassium voltage-gated channels during repolarization. A negative voltage means the inside of the fibre is negative compared with the outside. A negative current means the current is flowing through channels into the fibre.

Graph 1 shows the current through chloride channels and graph 2 shows the current through potassium channels.

Graph 1   Graph 2

[Source: Christopher W. Waters, Grigor Varuzhanyan, Robert J. Talmadge, and Andrew A. Voss, 'Huntington disease skeletal muscle is hyperexcitable owing to chloride and potassium channel dysfunction' (Proceedings of the National Academy of Sciences May 2013, 110 (22) 9160–9165; DOI: 10.1073/pnas.1220068110)]

(f) Identify if potassium channels in WT mice are open or closed at: **[1]**

(i) −60 mV

....

(ii) 20 mV.

....

Make sure you are looking at the graph with potassium channels not chloride channels.

(g) Deduce the relationship between voltage (mV) and the current flow through potassium channels in WT mice. **[1]**

....

....

(h) Using the data in both graphs, deduce whether HD increases or decreases current flow through chloride and potassium channels. **[2]**

....

....

....

....

Chloride channels are affected by the expression of a gene ClC-1 via translation of the mature mRNA Clcn1. Immature Clcn1 can be spliced differently to make mature mRNA with ($7a^+$) or without ($7a^-$) exon 7a.

Gel electrophoresis was carried out to identify two forms of the Clcn1 mature mRNA in control mice (WT) and mice with Huntington's disease (HD).

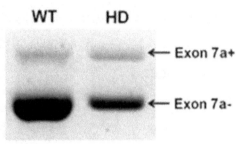

[Source: Christopher W. Waters, Grigor Varuzhanyan, Robert J. Talmadge, and Andrew A. Voss, 'Huntington disease skeletal muscle is hyperexcitable owing to chloride and potassium channel dysfunction' (Proceedings of the National Academy of Sciences May 2013, 110 (22) 9160-9165; DOI: 10.1073/pnas.1220068110)]

Look at the band thickness and use the text to identify the key, WT and HD.

(i) State which form of mRNA is required for normal chloride channel function. **[1]**

The study then investigated the percentage of mRNA that is 7a+ and the overall expression of the Clcn1 mRNA. The results are shown below.

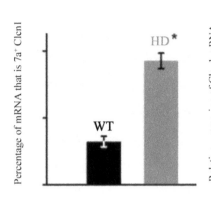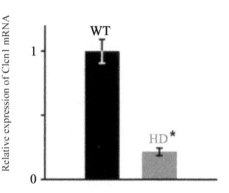

[Source: Christopher W. Waters, Grigor Varuzhanyan, Robert J. Talmadge, and Andrew A. Voss, 'Huntington disease skeletal muscle is hyperexcitable owing to chloride and potassium channel dysfunction' (Proceedings of the National Academy of Sciences May 2013, 110 (22) 9160-9165; DOI: 10.1073/pnas.1220068110)]

(j) Analyse the effect of the $7a^+$ form of Clcn1 mRNA on Huntington's disease. **[3]**

ANSWER ANALYSIS

Use the graph axis to help you. Ideally use manipulated data or orders of magnitude/ comparisons/ percentages rather than stating values.

(k) Based on the data and your biological knowledge, suggest causes for the prolonged involuntary skeletal muscle contraction in Huntington's disease. **[3]**

ANSWER ANALYSIS

Use information from all parts of the question for full marks. Which depolarizes more easily? – link to evidence. Which takes longer to repolarize? Why does it take longer to repolarize? Which channels are affected?

2. The image shows an embryonic stem cell.

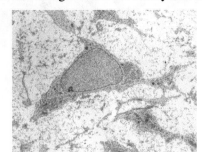

(a) Identify, with a reason, which stage of the cell cycle it is in. **[1]**

....

....

(b) State the characteristics of stem cells. **[2]**

....

....

....

....

(c) Explain the control of the cell cycle. **[3]**

....

....

....

....

....

....

(d) Outline the cause and therapeutic treatment of Stargardt's disease. **[2]**

....

....

....

....

3. (a) Define metabolism. **[1]**

....

....

The sketch shows the effect of changing substrate concentration on the rate of a reaction with and without an inhibitor.

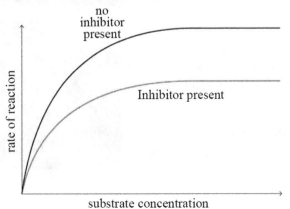

> The cell cycle consists of interphase and mitosis (prophase, metaphase, anaphase and telophase).

> Use terms unspecialized/specialized, undifferentiated/differentiated, totipotent, pluripotent or multipotent.

> What stages make up the cell cycle? How are cyclins involved? What are kinases?

> Where is the mutation and what is the treatment? It will be linked to the question so look at part b).

> Make sure you learn simple definitions – search the syllabus and make a list.

(b) Determine the type of inhibition. [1]

....

(c) Explain the effect of the inhibitor on the rate of reaction. [2]

....

....

....

....

You should recognize the effect of increasing substrate concentration on both a competitive and non-competitive inhibitor.

EXPLAIN

This is an explain question so explain where the inhibitor binds, how it affects the active site and whether it is reversible.

4. In *Zea mays* plants the gene for seed colour has the allele C for coloured seed and c for colourless seed. The gene for endosperm has the allele S for starchy and s for waxy endosperm.

A test cross was carried out between parents as follows:

	Parent 1	Parent 2
	Heterozygous	**Homozygous recessive**
Parent	Coloured starchy seeds X	Colourless waxy seeds
Genotype	$\dfrac{C \quad S}{c \quad s}$	$\dfrac{c \quad s}{c \quad s}$

Phenotype of offspring	Number
Coloured waxy kernels	58
Colourless waxy kernels	196
Coloured starchy kernels	192
Colourless starchy kernels	54

The outcome of the cross is below.

(a) State the type of inheritance shown. [1]

....

(b) Identify the four possible gametes for parent 1. [2]

....

....

....

....

....

Don't forget one chromosome must come from the homozygous recessive parent.

(c) Using the correct notation, identify the genotypes of the offspring which are recombinants. [1]

....

....

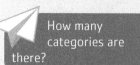

How many categories are there?

(d) A chi-squared (χ^2) value was calculated as shown to determine if the observed ratio (O) differed significantly from the expected Mendelian ratio Zeal (E). The chi-squared value was 116.57.

Determine if the observed ratio differed from the expected ratio and give reasons. [2]

$$\chi^2 = \sum \frac{(O-E)^2}{E} = 116.57$$

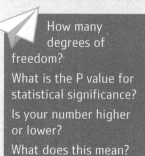

How many degrees of freedom?

What is the P value for statistical significance?

Is your number higher or lower?

What does this mean?

	P value (Probability)										
Degrees of freedom (df)	0.95	0.9	0.8	0.7	0.5	0.3	0.2	0.1	0.05	0.01	0.001
1	0.004	0.02	0.06	0.15	0.46	1.07	1.64	2.71	3.84	6.63	10.83
2	0.1	0.21	0.45	0.71	1.39	2.41	3.22	4.61	5.99	9.21	13.82
3	0.35	0.58	1.01	1.42	2.37	3.66	4.64	6.25	7.81	11.34	16.27
4	0.71	1.06	1.65	2.2	3.36	4.88	5.99	7.78	9.49	13.28	18.47
5	1.14	1.61	2.34	3	4.35	6.06	7.29	9.24	11.07	15.09	20.52
6	1.63	2.2	3.07	3.83	5.35	7.23	8.56	10.64	12.59	16.81	22.46

(e) Explain how recombinants are formed during meiosis. **[2]**

> Which stage of meiosis does recombination occur in? What is the effect of crossing over on allele combinations?

(f) *Zea mays* belongs to the plant phylum angiospermophytes. Distinguish between angiospermophytes and bryophytes. **[2]**

> You need two different direct comparisons. A table would be useful here.

Section B

Answer **two** questions from a choice of three. Up to one additional mark is available for the construction of your answers for each question.

5. Technology has made use of variation in the base sequence of genetic material.

(a) Distinguish between genes and short tandem repeats in DNA. [3]

ANSWER ANALYSIS

A table would help here. Think about sequences of bases – are they repetitive or not; do they code for a polypeptide? Are they unique in the genome? Are they translated?

(b) Outline how DNA profiles are made and give a use of a DNA profile. [4]

OUTLINE
Give a brief account or summary.

Use terms such as PCR, electrophoresis, fragments, and uses of DNA profiles.

(c) Explain how biochemical evidence supports evolution. [8]

Define evolution. Explain how DNA is universal. Why is this useful when comparing species? Describe the use of gene banks/ protein banks. Why are gene banks better than previous classification based on phenotypes? What is a molecular clock?

6. Increases in global travel can impact human health.

(a) Outline the role of immunoglobulins. **[3]**

....

....

....

....

....

Immunoglobulins are antibodies. What is the role of antibodies in the specific immune response? How do antibodies destroy pathogens?

💬 **OUTLINE**

Give a brief account or a summary.

(b) Explain the causes of jet lag on circadian rhythms in humans. **[6]**

....

....

....

....

....

....

....

....

....

....

....

Define circadian rhythm. What hormone is involved? Where is the hormone made and where does it act upon? What is the effect of the hormone? How does jet lag affect the hormone level? What is the consequence of the altered hormone level?

💬 **EXPLAIN**

Give a detailed account including the reasons or causes.

(c) Explain the way vaccination provides immunity. **[6]**

....

....

....

....

....

....

....

....

....

....

....

What is a vaccine? Which cells recognize the antigen? Explain antigen presentation and T cell and B cell involvement. Why are vaccines useful?

7. Amino acids are broken down by organisms and excreted as nitrogenous waste.

(a) Draw a labelled diagram showing the structure of a human kidney. **[3]**

> This is the visible structure you would see in a dissection rather than the detailed structure of a nephron.

(b) Compare and contrast the composition of blood in the renal artery and renal vein. **[4]**

> Composition refers to the components of the blood rather than the pressure or differences in the structure of the artery and vein. Think about the levels of proteins, oxygen, glucose, urea, salt, and water.

(c) Explain how insects eliminate nitrogenous waste. **[8]**

> Explain in detail how insects use Malpighian tubules, active transport, osmosis, haemolymph and uric acid.

Paper 3: Higher Level

- Set your timer for **1 hour and 15 minutes**
- The maximum mark for this examination paper is **[45 marks]**
- **Section A** – Answer ALL the questions
- **Section B** – Answer all of the questions in **ONE** of the options
- A calculator is required for this paper

Section A

1. A student investigated osmosis in 9 potato (*Solanum tuberosum*) cubes and 9 sweet potato (*Ipomoea batatas*) cubes in different concentrations of salt solution. They measured 1 gram of sweet potato and potato. The student then placed them into test tubes containing 20 cm^3 of various concentrations of salt solutions (NaCl). After 1 hour the final mass was recorded by removing the potato/sweet potato and putting it on the scales. The percentage change in mass was plotted against the concentration of salt solution for both potato and sweet potato.

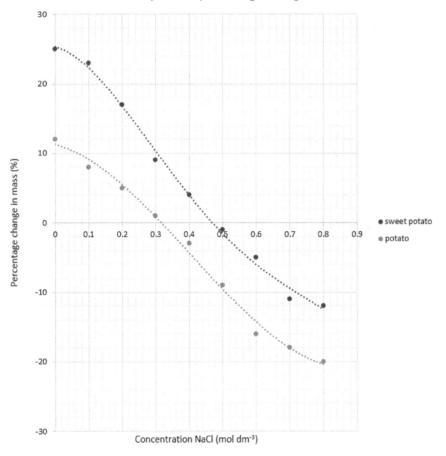

How salinity affects percentage change in mass

(a) State a controlled variable in the experiment. **[1]**

....

(b) Determine the isotonic point of potato and sweet potato. **[1]**

....

(c) Suggest the concentration where the potato is in a hypertonic solution and the sweet potato is in a hypotonic solution. **[1]**

....

(d) Calculate the osmolarity of the potato and sweet potato, giving the units. **[2]**

Potato ....

Sweet potato ....

> A controlled variable is a variable that is kept constant for a fair test.

> The isotonic point is where the potato neither gained or lost mass.

> Water moves from a hypotonic (dilute solution) to a hypertonic (concentrated solution). Osmosis is the movement of water from a dilute to a concentrated solution across a partially permeable membrane. If the tissue gains water by osmosis the solution is hypotonic (dilute), if the tissue loses water by osmosis the solution is hypertonic (concentrated).

> Osmolarity takes into account the total number of particles making up the solutes in the solution as well as the isotonic concentration.

> Osmolarity = molarity × *n*. *n* is the number of particles or ions present in the solution. For instance, a solution of NaCl will dissociate into two ions that are Na+ and Cl-.

(e) Identify a source of error in the procedure. [1]

....

....

What controlled variables is it difficult to keep constant?

Below is the nutritional information for potatoes and sweet potatoes.

Nutrient per 100 g	Potatoes	Sweet potatoes
Water (g)	**79**	77
Energy (kJ)	322	360
Protein (g)	2.0	1.6
Fat (g)	0.09	0.05
Carbohydrates (g)	17	20
Fibre (g)	2.2	3
Sugar (g)	0.78	4.18

(f) Deduce the reason for the difference in the isotonic point between potato and sweet potato. [1]

....

....

Look at the table and link the data showing a big difference in solute concentration to your knowledge of osmosis.

2. The image below shows a light micrograph image of broad bean (*Vicia faba*) root cells.

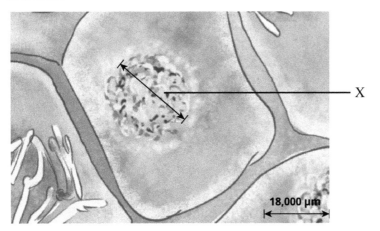

X

18,000 µm

Learn what the stages of mitosis look like under the microscope as well as in diagrams.

(a) Identify the stage of mitosis for cell X. [1]

X = ...

(b) Determine the actual diameter of the nucleus in cell X. [1]

Measure the length of the scale bar in mm. Convert to micrometres to get the image size. Read off the actual size from the scale bar in micrometres. Use the magnification formula.

....

A student counted the number of cells in different stages of the cell cycle in two different areas of a plant root.

Area X

Phase of cell cycle	Number of cells observed +/− 10
Interphase	40
Prophase	27
Metaphase	17
Anaphase	16
Telophase	6
Total	106

Area Y

Phase of cell cycle	Number of cells observed +/− 10
Interphase	188
Prophase	50
Metaphase	8
Anaphase	9
Telophase	1
Total	256

(c) Calculate the mitotic index for both areas X and Y. **[1]**

X = ...

Y = ...

The image shows an allium root tip × 10.
(1 – meristem quiescent centre; 2 – calyptrogen (live rootcap cells); 3 – rootcap; 4 – sloughed off dead rootcap cells; 5 – procambium)

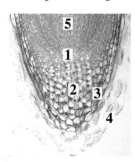

(d) Deduce, with a reason, which region of the root the data in area X was collected from. **[1]**

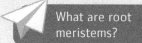
What are root meristems?

....

3. In 1952 Alfred Hershey and Martha Chase investigated whether protein or DNA was the genetic material. They infected bacterial cells with a virus (bacteriophage) containing either a radioactive sulphur coat or radioactive DNA. Below is a summary of the Hershey–Chase experiment.

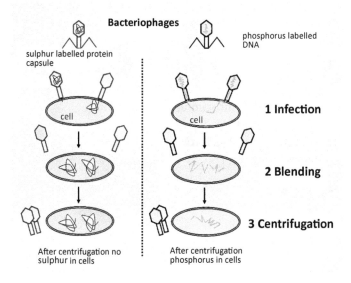

(a) Suggest why radioactive sulphur was used to label proteins and radioactive phosphorus used to label DNA. **[1]**

In the biochemistry unit – Where is P found? Where is S found?

....

....

(b) Deduce why radioactive carbon was not used. **[1]**

Why would C not differentiate between DNA and proteins?

....

....

(c) Explain the results of the experiment. **[2]**

Consider where the radioactivity was found and what this suggested.

....

....

....

....

Section B

Option A – Neurobiology and behaviour

1. The image shows a tissue slice from the brain of an adult macaque monkey (*Macaca mulatta*).

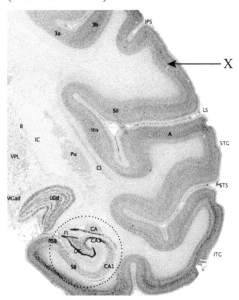

(a) State which part of the brain the outer layer labelled X represents. **[1]**

(b) Suggest a reason why the structure in this area of the brain differs in humans and macaque monkeys. **[1]**

A study investigated the amount of brain metabolism (measured as micromoles of glucose consumed per minute) used by rodents (*Glires*) and primates (*Primata*) as a function of the total number of neurons in the brain. The arrow represents humans.

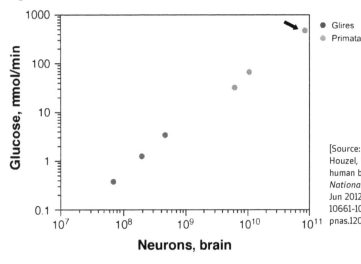

[Source: Suzana Herculano-Houzel, 'The not extraordinary human brain', *Proceedings of the National Academy of Sciences* Jun 2012, 109 (Supplement 1) 10661-10668; DOI: 10.1073/pnas.1201895109]

(c) Suggest a reason why primates have a higher glucose requirement than rodents. **[2]**

(d) Explain how memories are formed. **[3]**

....

....

....

....

....

....

What is memory? What is encoding and retrieval? What is neural pruning?

2. Below are the results of a study investigating the activation of the visual cortex after training with a soundscape. A soundscape is a way of representing a visual stimulus as a sound. I shows the baseline control with no training. II–IV show the results after training following exposure to II sound (auditory stimulus) on the auditory cortex, III visual stimulus on the visual cortex and IV auditory stimulus on the visual cortex.

Baseline activation — No training | Auditory cortex activation (auditory encoded stimulus) Post training | Visual cortex activation (visual stimulus) Post training | Visual cortex activation (auditory encoded stimulus) Post training

[Source: Lotfi Merabet, Dorothe Poggel, William Stern, Ela Bhatt, Christopher Hemond, Sara Maguire, Peter Meijer and Alvaro Pascual-Leone, 'Activation of visual cortex using crossmodal retinotopic mapping', www.seeingwithsound.com/hbm2008.html]

(a) Deduce the effect of both visual and sound stimuli on the activation of the visual cortex post training. **[1]**

....

....

Use the information to state how sound affects the activation of the visual cortex.

(b) Suggest the advantages of neural plasticity to a visually impaired person. **[2]**

....

....

....

....

Why won't a hearing aid work? How do the cochlea implants process sound?

(c) Sometimes people who cannot hear have a cochlear implant. Describe how the cochlea implant allows the user to hear sound. **[4]**

....

....

....

....

....

....

....

....

3. A student investigated the effect of four conditions on the behaviour of 25 woodlice (*Porcellio scaber*). Woodlice breathe through gills and feed on decaying organic plant matter such as dead leaves. The choice chamber had four different conditions: dry and dark, dry and light, damp and dark, and damp and light. The results are shown below.

Results

Condition	Number of woodlice in each condition					
	1	2	3	4	5	total
Dry dark	4	5	5	6	7	27
Dry light	2	3	3	3	2	13
Damp dark	16	13	11	12	12	64
Damp light	3	4	6	4	4	21

(a) State what type of behaviour the woodlice were showing in response to the light? **[1]**

....

....

Positive/negative; geotaxis/ phototaxis?

(b) Define innate behaviour. **[1]**

....

....

The chi-squared table below was used to determine if the difference is statistically significant.

$$\chi^2 = \sum \frac{(O - E)^2}{E}$$

χ^2 = the test statistic $\sum$ = the sum of

O = Observed frequencies   E = Expected frequencies

df	P = 0.05	P = 0.01	P = 0.001
1	3.84	6.64	10.83
2	5.99	9.21	13.82
3	7.82	11.35	16.27
4	9.49	13.28	18.47
5	11.07	15.09	20.52

The value of chi^2 (χ^2) is 67.4

(c) Deduce if the woodlice have a preference for a habitat. **[3]**

....

....

....

....

....

....

Consider:
• How many degrees of freedom?
• What is the critical value?
• How can you interpret your value?

4. Common vampire bats (*Desmodus rotundus*) are nocturnal and feed on blood. They starve and die if they don't feed for 70 hours. Successful hunters regurgitate blood to feed starving bats. A study investigated 20 tagged vampire bats over a two-year period to investigate the reason for sharing of food.

11 males and 9 females out of 25 were fasted for 24 hours then returned to their cage. Social interactions were recorded for 2 hours.

The graphs below compare the amount of food donated with the amount of food received, the allogrooming received (reciprocal grooming) and the relatedness (how genetically close the bats are). The amount donated and received was converted to a z-score so results could be compared – a higher z-score means a higher amount donated/received.

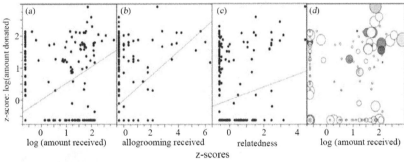

[Source: Carter Gerald G. and Wilkinson Gerald S., 2013, 'Food sharing in vampire bats: reciprocal help predicts donations more than relatedness or harassment', Proc. R. Soc. B.2802012257320122573, http://doi.org/10.1098/rspb.2012.2573]

(a) Use the graphs to determine if relatedness between individuals is the strongest motivator for sharing food. **[2]**

Look at the graphs and decide if genetics is the strongest motivator. Use evidence from the graphs to back up your answer.

(b) Outline the type of behaviour the bats are showing and how it can help the survival of the bat colony. **[2]**

(c) The study observed males did not feed other males. Suggest why. **[1]**

Why do males compete with each other?

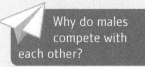

5. Explain Skinner's experiment into operant conditioning. **[6]**

....

....

....

....

....

....

....

....

....

....

....

Skinner tested positive and negative reinforcement in rats. What did he do? What were the results? Use the terms learned, reinforcement, behaviour and operant response.

Option B – Biotechnology and bioinformatics

1. A study tracked tumour growth in control mice (vehicle) and experimental mice on an anti-cancer drug sorafenib at a dose of 40 mg/kg. The animal tumour model were hairless *Mus musculus* mice that had been bred to have no immune systems.

Tumour-inducing inoculation was given 2 weeks before the study. Image I shows tumour growth was tracked over 10 days using a fluorescent dye-tagged protein with a CT imaging system. The tumour is highlighted with a black circle. Image II shows the same tumour using a PET scan.

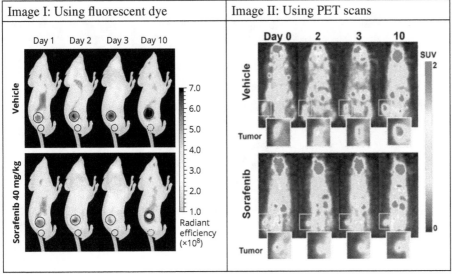

Image I: Using fluorescent dye	Image II: Using PET scans

[Source: Tseng, J.C., Narayanan, N., Ho, G., Groves, K., Delaney, J., et al. (2017) 'Fluorescence imaging of bombesin and transferrin receptor expression is comparable to 18F-FDG PET in early detection of sorafenib-induced changes in tumor metabolism'. PLOS ONE 12(8): e0182689. https://doi.org/10.1371/journal.pone.0182689]

(a) Evaluate advantages and disadvantages of using this model of mice in the study. **[4]**

....

....

....

....

....

....

....

....

What are the advantages and disadvantages of using the mice as models for human beings? What is useful about this particular mouse model? Consider animal testing. Look at the information at the start of the question for help.

Look at the resolution of the two images.

(b) Determine, with a reason, whether fluorescent dye or a PET scan is better at tracking tumours. **[1]**

....

....

(c) State a suitable protein to attach to the fluorescent dye to produce these images. **[1]**

....

(d) Deduce, with a reason, which mouse has a greater size of tumour after 10 days. **[1]**

....

....

(e) Evaluate the use of the drug sorafenib at a dose of 40 mg/kg as an anti-cancer drug. [4]

....

....

....

....

....

....

....

Use the information in the study to state what you see for a 40 mg dose compared with the control 'vehicle' mouse. State if it is effective, when it is effective and if differences between the control and 40 mg mouse are significant. Use evidence/data to back it up and justify your reasoning with an interpretation.

2. The image shows the heel prick test collecting blood from a two-week-old infant and screening it for phenylketonuria, or PKU.

(a) Outline the importance of the early detection of PKU. [3]

....

....

....

....

....

....

How do abnormal metabolites affect function? What is the result if PKU is not detected? If PKU is detected how is it treated and what is the impact on life?

(b) Cystic fibrosis is a recessive autosomal inherited disease resulting in breathing difficulties. Cystic fibrosis can be detected by the heel prick test for high levels of the metabolite trypsin or an ELISA test for the protein trypsin using immobilized antibodies. Outline how the ELISA test detects abnormally high amounts of the metabolite trypsin in cystic fibrosis. [3]

....

....

....

....

....

....

What does ELISA stand for? What is ELISA used on? How does ELISA work? How are monoclonal antibodies involved? How are changes detected?

3. The image shows a duck from an oil spill that released about 58,000 gallons of oil spilled from a South Korea-bound container ship when it struck a tower supporting the San Francisco-Oakland Bay Bridge in dense fog on 11 July 2007. Bioremediation is one method used to treat oil spills.

(a) Define bioremediation. [1]

....

....

(b) Outline the use of bacteria in the bioremediation of oil spills. **[5]**

What are oil spills? What problem is caused by hydrocarbons? Which bacteria break down hydrocarbons? Compare physical and chemical methods of bioremediation.

4. Below is a section of information comparing the nucleotide sequence from the gene for actin in the dingo *Canis lupis dingo* with the dog *Canis lupis familiaris* (Subject 1).

PREDICTED: Canis lupus dingo filamin A (FLNA), transcript variant X1, mRNA

Sequence ID: XM_025460575.1 Length: **8497** Number of Matches: **1**

Range 1: 1 to 8497 GenBank Graphics   ▼ Next Match ▲ Previous

Score	Expect	Identities	Gaps	Strand
15692 bits(8497)	0.0	8497/8497(100%)	0/8497(0%)	Plus/Plus

```
Query  1   CGTGGAGGCGCGCGGCGCTCAGCGGACGCCAGCGGAACCCCGAGGCGCCGGGCGCGGGCG  60
           ||||||||||||||||||||||||||||||||||||||||||||||||||||||||||||
Sbjct  1   CGTGGAGGCGCGCGGCGCTCAGCGGACGCCAGCGGAACCCCGAGGCGCCGGGCGCGGGCG  60

Query  61  CAGAGGCGACCCCGGCGCCACCCCCGCAGTCGCGGGCCGCTCGCAGGGACAGCAGCACAA  120
           ||||||||||||||||||||||||||||||||||||||||||||||||||||||||||||
Sbjct  61  CAGAGGCGACCCCGGCGCCACCCCCGCAGTCGCGGGCCGCTCGCAGGGACAGCAGCACAA  120
```

[Source: National Library of Medicine (NLM)]

(a) State the length of the nucleotide. **[1]**

How many base pairs long is the nucleotide?

(b) Determine, with a reason, if the nucleotide search was DNA or RNA. **[1]**

Which bases are present in the genetic material?

(c) Use the information to suggest why there is debate as to whether dingos and dogs are the same species or different species. **[2]**

Is the DNA similar? What does this suggest? Remember we are only looking at one gene here.

How could the degenerate nature of DNA disguise mutations?

(d) Outline whether nucleotide sequences or amino acid sequences are likely to give more reliable data. **[1]**

Actin is a protein found in muscles!

(e) Explain whether actin could be used to develop evolutionary relationships between plant phyla. **[2]**

Option C – Ecology and conservation

1. Barnacles are non-motile filter feeders that cement themselves to rocks and feed on plankton and detritus when immersed in sea water. When the tide recedes, they are exposed to the atmosphere (emersed) and close their shell to avoid desiccation. The kite chart below shows the distribution of three species of barnacle on an exposed rocky seashore in Wales. The data were collected from mid-shore (lower mid-shore is 2.5 m, upper mid-shore is 5.2 m and upper shore is 6 m).

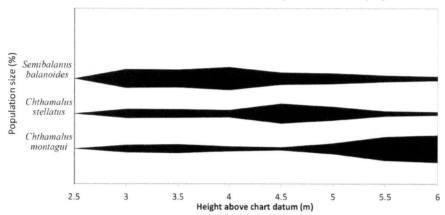

(a) Outline the method used to collect the data. **[2]**

(b) Suggest why numbers of all barnacles are low at 2.5 m above chart datum. **[1]**

(c) Deduce, with a reason, the species least tolerant to exposure and desiccation. **[1]**

(d) Explain the zonation of the barnacles using the principle of competitive exclusion. **[3]**

(e) Outline the difference between the fundamental and realized niche of *Chthamalus montagui*. **[2]**

How are data collected along an abiotic gradient in ecology?

At low shore there are lots of algae and limpets – how could this affect the barnacles?

At low shore the tide covers the barnacles for longer.

What is the principle of competitive exclusion? Use the kite diagram to deduce which species is outcompeting which at each height, and why.

The fundamental niche is all the environmental conditions the species could possibly live in, whereas the realized niche is where it actually lives due to constraints such as competition and predation.

2. The web below shows interactions between species in the North Pacific Ocean. The arrows represent known (solid arrows) or suspected (dashed arrows) interactions between species. Interactions can have top-down or bottom-up effects, direct or indirect effects, and positive or negative effects on the populations.

> **Key**
>
> Black lines represent top-down effects
> Grey lines represent bottom-up effects

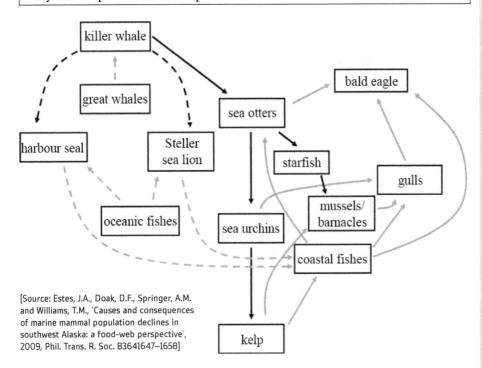

[Source: Estes, J.A., Doak, D.F., Springer, A.M. and Williams, T.M., 'Causes and consequences of marine mammal population declines in southwest Alaska: a food-web perspective', 2009, Phil. Trans. R. Soc. B3641647–1658]

(a) Distinguish between top-down and bottom-up limiting factors. [1]

....

....

> What is the difference between top-down and bottom-up limiting factors?

(b) Identify the role of kelp in the food web. [1]

....

....

> What part of the food chain/web does kelp occupy?

(c) Suggest how an increase in the kelp population could affect other populations in the web above. [2]

....

....

....

....

> What will increase/decrease and why?

> What is a keystone species? What is the effect on an ecosystem if the keystone species is removed?

(d) The otter is a keystone species. Outline the role of a keystone species. [1]

....

....

(e) Suggest, with a reason, what would happen to kelp if the number of sea otters decreased. [1]

....

....

> Use the interaction between kelp and otters.

3. The chart shows boxplots of the mean net primary productivity of different biomes arranged in order of the mean biome. Bars within the boxes represent median values. The bottom and top of the box represent the 25th and 75th percentile, respectively. The bars outside the box represent the 10th and 90th percentiles. Open circles represent outliers.

Key			
DES	shrublands/deserts	TRS	tropical savannas
TUN	tundra	TMD	temperate deciduous forests
BW	boreal/taiga woodlands	TMM	temperate mixed forests
TMS	temperate savannas	TRD	tropical deciduous forests
BF	boreal/taiga forests	TMB	temperate broad-leaved
GRS	grasslands		evergreen forests
XFW	xeromorphic woodlands	TRE	tropical rainforests
TMC	temperate coniferous forests		

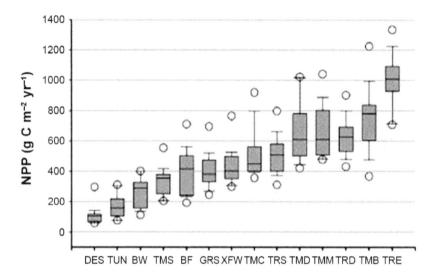

[Source: Kicklighter, D.W., Bondeau, A., Schloss, A.L., Kaduk, J., Mcguire, A.D. and Intercomparison, T.P.O.T.P.N.M., 'Comparing global models of terrestrial net primary productivity (NPP): global pattern and differentiation by major biomes.' Global Change Biology, 1999, 5: 16-24]

(a) Compare and contrast the mean net primary productivity in desert and tropical rainforests. **[2]**

....

....

....

....

> Read off the mean from the DES and the mean from the TRE.
>
> How many times higher is the TRE than the DES? Consider the variation in the two areas.

(b) Suggest a reason for the difference in net primary productivity in desert and tropical rainforest biomes. **[1]**

....

....

(c) Suggest, with a reason, which biome has longer food chains. **[1]**

....

....

> Remember only 10% of energy is passed on at each trophic level!

4. The table below shows the genus of coral found at different depths at the Marine Park in Sulaweisi, Indonesia, Kaledupa.

Genus of coral	5 m depth		10 m depth	
	n	n(n − 1)	n	n(n − 1)
Acropora	15	210	25	600
Astreopora	3	6	12	132
Cyphastrea	0	0	3	6

[Source: Adapted from Table 2: David J. Smith, 'Marine Report: Marine biodiversity and ecology of the Wakatobi Marine National Park', Southeast Sulawesi (August 2003) p.62 (https://cdn.yello.link/opwall/files/2017/11/Opwall-Indonesia-Wakatobi-Marine-Science-Report-2003.pdf)]

Genus of coral	5 m depth		10 m depth	
	n	n(n − 1)	n	n(n − 1)
Diploastrea	0	0	3	6
Euphyllia	4	12	1	0
Favia	14	182	10	90
Favites	8	56	12	132
Fungia	15	210	6	30
Galaxea	4	12	4	12
Goniapora	2	2	2	2
Goniasrea	24	552	26	650
Heliofungia	1	0	1	0
Leptoseris	3	6	16	240
Montipora	5	20	8	56
Mycedium	0	0	2	2
Pachyseris	0	0	0	0
Physogyra	0	0	4	12
Plerogyra	4	12	10	90
Pocillopora	8	56	17	272
Porites	45	1980	31	930
Sarcophyton	33	1056	23	506
Seriatopora	4	12	16	240
Stylophora	18	306	27	702
Symphyllia	9	72	3	6
Turbinaria	0	0	2	2
Total	**219**	**4762**	**264**	**4718**
Simpson's reciprocal index		**10.02**		

(a) One way of measuring diversity is species richness. State the species richness for 5 m and 10 m. **[1]**

5 m: .... 10 m: ....

> Species richness refers to how many different species are present.

Simpson's reciprocal diversity index, D, is a more accurate method used to calculate the diversity of an ecosystem as it takes into account both species richness and evenness.

(b) Define the term evenness. **[1]**

....

....

> Learn definitions. Evenness refers to how close the abundance of each species is in the environment.

(c) Use the formula below and the totals in the table to calculate Simpson's reciprocal index for 10 m. **[1]**

$$D = \frac{N(N-1)}{\sum n(n-1)}$$

....

....

> N = total number of organisms of all species
> n = total number of organisms of a particular species.

(d) Deduce which depth has higher biodiversity. **[1]**

....

....

> The higher the value, the greater the diversity. (The maximum possible value is the number of species present.)

(e) Suggest one biotic factor that may change as depth increases. **[1]**

....

....

> What are biotic factors? How do biotic interactions affect population sizes?

(f) Distinguish between rainforest and desert biomes using Gersmehl diagrams. Refer to climate, nutrient stores and nutrient flows. **[6]**

Draw Gersmehl diagrams to help you. Don't forget the thickness of the arrows is important for transfers between stores, and the size of the circles is important for stores.

Option D – Human physiology

1. The electrocardiogram below shows a heart trace of a patient with anorexia.

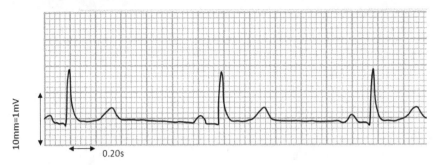

10mm=1mV

0.20s

(a) Label a P wave on the electrocardiogram above. [1]

(b) Calculate the heart rate in the electrocardiogram in BPM. [1]

....

(c) Anorexic people have slow heart rates to conserve energy. The heart rate is slowed down by nervous control. State the name of the tissue targeted by nervous control to lower the heart rate. [1]

....

The table below shows information about the heart for 91 subjects with anorexia nervosa and 62 control subjects. Values shown are means +/- standard deviation.

	Patients with anorexia nervosa (n = 91)	Control subjects (n = 62)
Heart rate (beats/min)	55.6 ± 11.4	72.4 ± 12.3
Stroke volume (mL/beat)	50.7 ± 13.3	± 13.3

$\bar{x} \pm$ SD.

(d) Calculate the cardiac output of the two groups, with units. [2]

....

The graph below shows the difference in the mass of the left ventricle between the groups. Error bars represent one standard deviation from the mean.

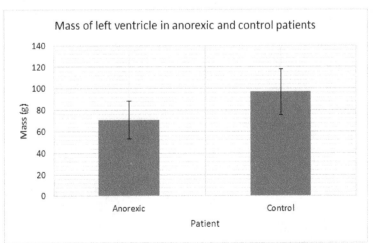

[Source: Adapted from: Carmela Romano et al., 'Reduced hemodynamic load and cardiac hypotrophy in patients with anorexia nervosa', *The American Journal of Clinical Nutrition*, Volume 77, Issue 2, February 2003, Pages 308–312]

(e) Calculate the percentage difference in the mass of the left ventricle. **[1]**

Percentage difference = (change/original) × 100

....

(f) Suggest a reason for the lower mass of the left ventricle in anorexic patients. **[1]**

....

....

The table below shows the difference in blood pressure between the two groups.

Blood pressure (mmHg)	Patients with anorexia nervosa (n = 91)	Control subjects (n = 62)
Systolic	90.8 ± 9.9	110.9 ± 10.8
Diastolic	62.1 ± 7.6	70.3 ± 8.9
Mean	71.7 ± 7.7	83.8 ± 8.2

$\bar{x} \pm$ SD.

[Source: Extract from Table 1 in Carmela Romano et al., 'Reduced hemodynamic load and cardiac hypotrophy in patients with anorexia nervosa', *The American Journal of Clinical Nutrition*, Volume 77, Issue 2, February 2003, Pages 308–312]

(g) Use the information to suggest the effect of anorexia on blood pressure. **[2]**

....

....

....

....

- Systolic pressure represents the pressure in the artery when the heart is contracting (systole).
- Diastolic pressure represents the pressure in the artery when the heart is relaxed (diastole).

2. The graph shows the oxygen dissociation curve for hemoglobin in both maternal and fetal blood.

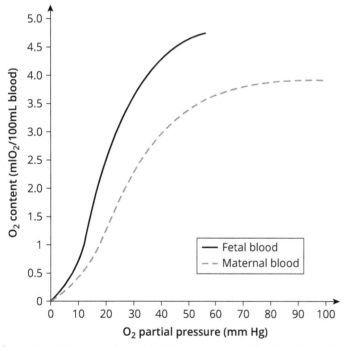

[Source: Parer, J.T., ed. 'Uteroplacental physiology and exchange' in *Handbook of Fetal Heart Rate Monitoring*. Philadelphia, PA: WB Saunders; 1997:40.]

(a) When the placental maternal blood has a partial pressure of 30 mmHg, calculate the percentage increase in the oxygen content of maternal and fetal blood. **[1]**

Draw a line up and across from 30 mmHg and read off the values for oxygen content.

Percentage increase = (change/original) × 100.

....

(b) Outline the reason for the difference in the oxygen dissociation curves
of the maternal and fetal hemoglobin. [2]

....

....

....

....

(c) Outline the structure of saturated oxy-hemoglobin. [1]

....

....

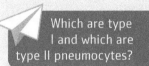

Which hemoglobin has a higher affinity for oxygen? Which way will oxygen therefore diffuse?

How many polypeptide chains are bound to how many heme groups? How many oxygen molecules are bound to each heme group?

3. The light micrograph below shows a cross-section of an alveolus wall ×400.

[Source: www.dartmouth.edu]

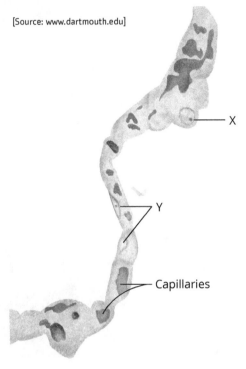

X

Y

Capillaries

(a) Identify cells X and Y. [1]

X = ...   Y = ...

Which are type I and which are type II pneumocytes?

The light micrograph below is taken from a diseased lung of a smoker.

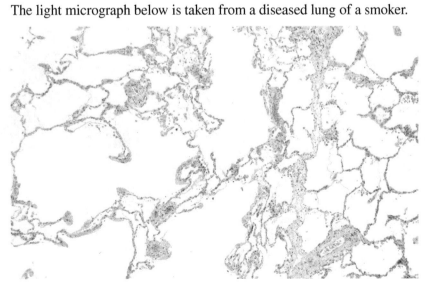

(b) Deduce, with a reason, the name of the disease. [1]

....

....

(c) Outline the causes and treatment of this disease. **[4]**

....

....

....

....

....

....

....

....

4. The image below shows part of a liver lobule.

(a) Identify X. **[1]**

....

(b) Outline causes and consequences of jaundice. **[4]**

....

....

....

....

....

....

....

What chemical causes jaundice? Why is there an increase of this chemical in the blood? How does it affect the skin and the feces?

(c) Outline the breakdown of erythrocytes. **[6]**

....

....

....

....

....

....

....

....

....

....

What does hemoglobin in red blood cells get broken down into? What happens to these products?

Set C

This set of papers has no additional help in the margins. There is a space to write notes so you can plan what you are going to write if needed.

Paper 1: Higher Level

- Set your timer for **1 hour**
- Each question is worth **[1] mark**
- The maximum mark for this examination paper is **[40 marks]**
- Answer ALL the questions

1. The electron micrograph below shows a eukaryotic organelle.

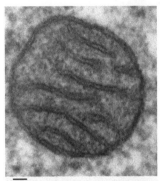

50 nm   08LungTEM

What is the magnification of the image? [1]

☐ A. × 1,500
☐ B. × 150,000
☐ C. × 15,000
☐ D. × 1,500,000

2. The diagram below shows a molecule. Identify the correct name of the molecule and the parts labelled X, Y and Z. [1]

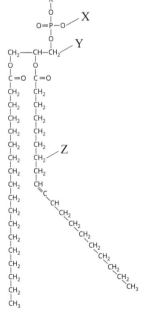

		Molecule name	X	Y	Z
☐	A.	Phospholipid	Glycerol	Protein	Fatty acid tail
☐	B.	Phospholipid	Phosphate	Glycerol	Fatty acid tail
☐	C.	Glycoprotein	Glycerol	Glycerol	Amino acid tail
☐	D.	Glycoprotein	Phosphate	Protein	Amino acid tail

NOTES

3. The diagram below shows two molecules. Identify the correct name of the molecules. [1]

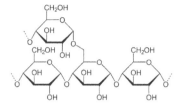

		I	II
☐	A.	α-glucose	Cellulose
☐	B.	Cellulose	β-glucose
☐	C.	Amylose	Amylopectin
☐	D.	Amylopectin	Amylose

4. The image below shows the protein hemoglobin. It is made up of four polypeptides and four heme groups. Which statement best describes structure X? [1]

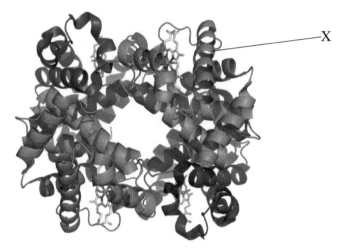

☐ A. Primary structure

☐ B. Alpha helix

☐ C. Beta pleated sheet

☐ D. Heme prosthetic group

5. A transect was set up to investigate how the population of a species changed in a woodland. Which statement could be concluded from a chi-squared test? [1]

☐ A. The population size increases as the distance into the woodland increases

☐ B. There is an association between the distance into the woodland and the species distribution

☐ C. There is a significant difference between the average number of species recorded at two different distances along the transect

☐ D. The increase in population size is directly proportional with the increase in distance into a woodland

6. Which theory did Meselson and Stahl obtain evidence to support? [1]

☐ A. Vitalism

☐ B. Spontaneous creation

☐ C. Cell theory

☐ D. Semi-conservative replication

NOTES

7. Which statement about the location of respiration is correct? **[1]**

	Glycolysis	**Krebs cycle**	**Electron transport chain**
☐ A.	Cytoplasm	Thylakoid	Cristae
☐ B.	Matrix	Matrix	Stroma
☐ C.	Cytoplasm	Matrix	Cristae
☐ D.	Matrix	Thylakoid	Stroma

8. Which of the following is a product of the link reaction? **[1]**

I. Pyruvate

II. Reduced NAD

III. Carbon Dioxide

IV. ATP

☐ A. I and II

☐ B. II and III

☐ C. I, II and III

☐ D. I, II, III and IV

9. Where is chlorophyll found in chloroplasts? **[1]**

☐ A. Stroma

☐ B. Matrix

☐ C. Thylakoid membranes

☐ D. Cristae

10. Which statement about surface area to volume ratio is true? **[1]**

☐ A. As cells increase in size their surface area to volume ratio increases

☐ B. Smaller cells have a higher surface area to volume ratio than larger cells

☐ C. A small surface area to volume ratio is needed to allow fast diffusion of oxygen into cells

☐ D. A small surface area to volume ratio releases less heat

11. The diagram summarizes respiration. What process is happening at I, II and III? **[1]**

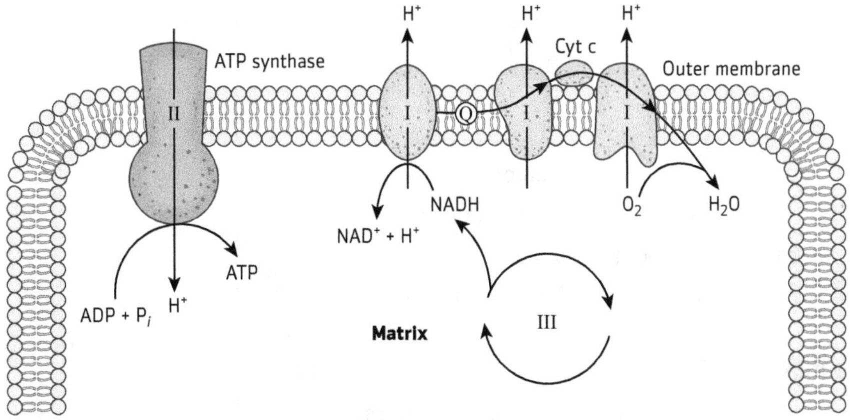

	I	**II**	**III**
☐ A.	Active transport of protons	Facilitated diffusion of protons	Krebs cycle
☐ B.	Facilitated diffusion of protons	Active transport of protons	Krebs cycle
☐ C.	Active transport of protons	Facilitated diffusion of protons	Calvin cycle
☐ D.	Facilitated diffusion of protons	Active transport of protons	Calvin cycle

NOTES

12. What are Okazaki fragments? **[1]**

☐ A. Short sections of RNA made in transcription

☐ B. Short sections of DNA made in DNA replication

☐ C. The name given to the primers removed from the lagging strand

☐ D. The sections of DNA used in gel electrophoresis

13. What is the correct mechanism of voltage-gated channels that generate an action potential? **[1]**

☐ A. Na^+ moves into the axon when a threshold potential is reached by facilitated diffusion

☐ B. Na^+ moves out of the axon when a threshold potential is reached by facilitated diffusion

☐ C. Na^+ moves into the axon when a threshold potential is reached by active transport

☐ D. Na^+ moves out of the axon when a threshold potential is reached by active transport

14. What processes do **not** occur in the placenta? **[1]**

☐ A. Small molecules such as glucose, oxygen and carbon dioxide diffuse across it

☐ B. Oestrogen is made and released into the mother's blood

☐ C. Antibodies pass from the baby to the mother

☐ D. The mother's blood mixes with the baby's blood

15. The images show a muscle cell. Identify structures X, Y and Z. **[1]**

		X	Y	Z
☐	A.	Myosin filament	Actin filament	Sarcomere
☐	B.	Sarcomere	Z band	Actin filament
☐	C.	Actin filament	Myosin filament	Z band
☐	D.	Z band	Sarcomere	Myosin filament

16. What is the correct sequence for making a DNA profile? **[1]**

☐ A. PCR → restriction enzyme → gel electrophoresis → autoradiography

☐ B. Restriction enzyme → PCR → gel electrophoresis → autoradiography

☐ C. Gel electrophoresis → autoradiography → PCR → restriction enzyme

☐ D. Autoradiography → PCR → restriction enzyme → gel electrophoresis

17. Which enzymes from DNA replication are correctly described? **[1]**

☐ A. DNA gyrase relieves the tension in the DNA

☐ B. DNA helicase breaks hydrogen bonds

☐ C. DNA polymerase III adds nucleotides in a 3 to 5 carbon direction

☐ D. DNA ligase joins the gaps between Okazaki fragments on the sense strand

NOTES

18. A parent organism with an unknown genotype is mated in a test cross. Approximately 50% of the offspring have the same phenotype as the unknown parent. What can be concluded? **[1]**

☐ A. The unknown parent is homozygous recessive for the trait

☐ B. The unknown parent is homozygous dominant for the trait

☐ C. The unknown parent is heterozygous recessive for the trait

☐ D. The unknown parent is co-dominant for the trait

19. In 1986 the Chernobyl nuclear accident sent out a large amount of the radioactive isotope of iodine into the environment.

The graph below shows the number of cases of thyroid cancer per 100,000 people in the following years.

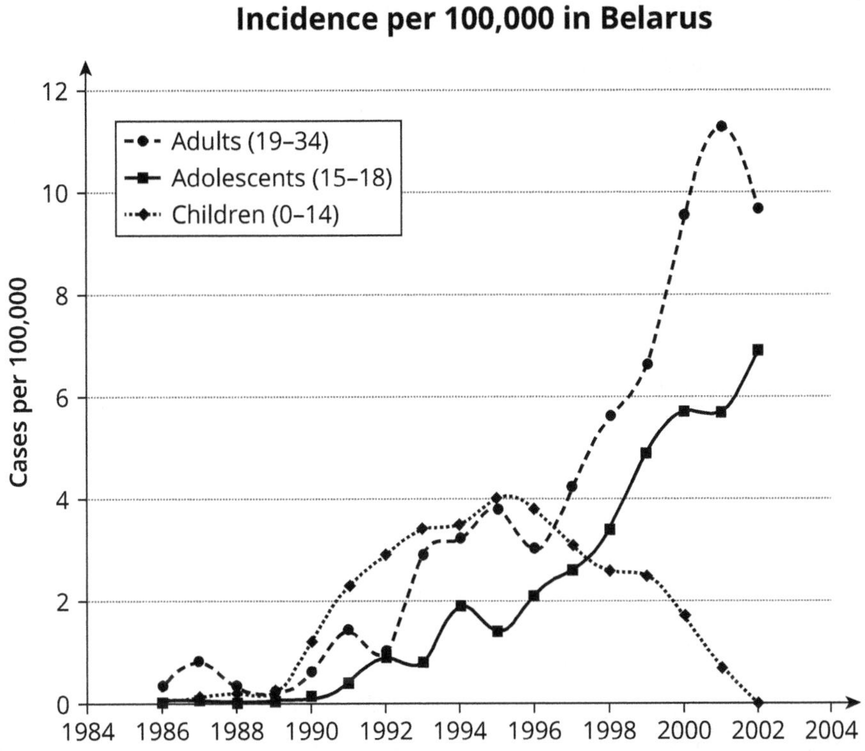

Incidence per 100,000 in Belarus

What conclusions can be made from this graph? **[1]**

☐ A. Adults had the greatest exposure to iodine ^{131}I

☐ B. All the children affected died by 2002

☐ C. A higher number of adults developed cancer than any other age group

☐ D. Chernobyl caused all the cases of cancer

20. *Euglena* is a microscopic protozoan that can be found in freshwater and saltwater environments. It uses its chloroplasts for photosynthesis, but also ingests dead organic matter by endocytosis. What is its mode of nutrition? **[1]**

☐ A. Autotroph only

☐ B. Heterotroph only

☐ C. Autotroph and detritivore

☐ D. Heterotroph and saprotroph

NOTES

21. Who discovered the circulation of blood? [1]

☐ A. Singer and Nicolson

☐ B. William Harvey

☐ C. Florey and Chain

☐ D. Louis Pasteur

22. Below is an image of the male reproductive system.
What are the correct labels for I, II and III? [1]

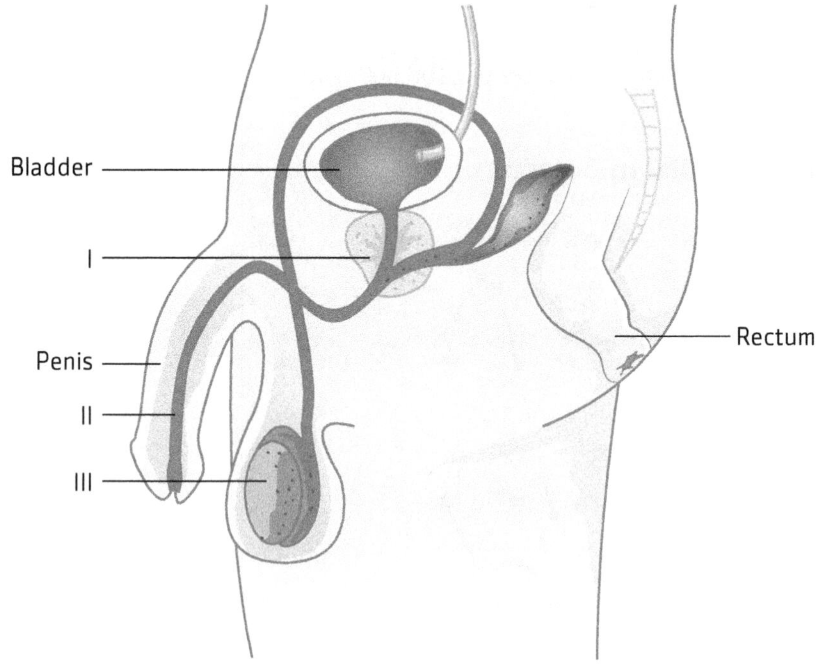

	I	II	III
☐ A.	Epididymis	Ureter	Testicle
☐ B.	Prostate gland	Urethra	Testis
☐ C.	Epididymis	Urethra	Testicle
☐ D.	Prostate gland	Ureter	Testis

23. Which of the following can affect gene expression?

I. Methylation

II. Acetylation

III. Phosphorylation [1]

☐ A. I only

☐ B. I and II only

☐ C. II and III only

☐ D. I, II and III

24. Which are features of tRNA?

I. Triplet of bases called an anticodon

II. Attachment site for amino acids

III. A pairs with T and C pairs with G by complementary base pairing [1]

☐ A. I only

☐ B. I and II only

☐ C. II and III only

☐ D. I, II and III

NOTES

25. What is the definition of the primary structure of a protein? **[1]**

☐ A. The three-dimensional conformation of the protein

☐ B. The number and sequence of amino acids in the polypeptide chain

☐ C. More than one polypeptide chain in the protein

☐ D. A non-protein prosthetic group attached

26. Which of the following statements about collagen are true?

I. Collagen is made up of three peptide chains

II. Collagen has ionic bonds between the R-groups of the amino acids in the peptide chains

III. Collagen is a quaternary protein **[1]**

☐ A. I only

☐ B. I and II only

☐ C. I and III only

☐ D. I, II and III

27. In fruit flies (*Drosophila melanogaster*) the allele for wild-type body colour is dominant to black body, and the allele for normal wings is dominant to vestigial wings.

Flies homozygous for wild-type body colour and normal wings were crossed with flies with black body and vestigial wings. 1,000 of the offspring survived to maturity and the number of offspring with each phenotype was recorded in the table below.

Phenotype of offspring	Frequency of offspring
Wild-type body colour, normal wings	460
Wild-type body colour, vestigial wings	50
Black colour body, normal wings	50
Black colour body, vestigial wings	440

What conclusion can be made from this result? **[1]**

☐ A. The genes are not on the same chromosome

☐ B. Polygenic inheritance has occurred

☐ C. The genes are linked

☐ D. The parental cross was not a heterozygote and a homozygote

28. In maize (*Zea mays*) the allele for purple colour kernels is dominant to the allele for yellow. The allele for smooth kernels is dominant to wrinkled.

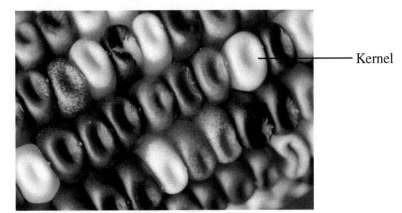 ———— Kernel

Corn homozygous for purple colour kernels (P) and smooth texture (S) was crossed with corn homozygous for yellow kernels (p) and wrinkly texture (s).

100 kernels in the offspring were counted and the number of offspring with each phenotype was recorded in the table on the following page.

NOTES

Phenotype of offspring	Frequency of offspring
Purple colour, smooth texture	56
Purple colour, wrinkly texture	19
Yellow colour, smooth texture	19
Yellow colour, wrinkly texture	6

What statistical test could be done to determine if the observed results differ significantly from the expected 9:3:3:1? **[1]**

☐　A.　Mean

☐　B.　Chi-squared test

☐　C.　Correlation

☐　D.　Standard deviation

29. The genes for body colour and wing type in fruit flies (*Drosophila melanogaster*) are linked.

Body colour gene: The allele for wild-type body is B and the allele for black body is b.

Wing type: The allele for normal wings is Vg and the allele for vestigial wings is vg.

The phenotypes are shown below.

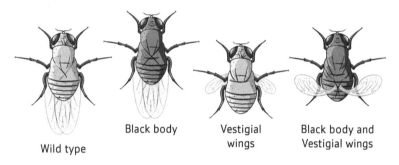

Wild type　　　Black body　　　Vestigial wings　　　Black body and Vestigial wings

If a heterozygous wild-type fly with normal wings is crossed with a black body fly with vestigial wings what genotype could be formed? **[1]**

$$\frac{B \quad Vg}{b \quad vg} \qquad \frac{b \quad vg}{b \quad vg}$$

☐　A.　$\dfrac{b \quad Vg}{b \quad vg}$

☐　B.　$\dfrac{b \quad Vg}{B \quad vg}$

☐　C.　$\dfrac{b \quad Vg}{b \quad Vg}$

☐　D.　$\dfrac{B \quad Vg}{B \quad Vg}$

NOTES

30. The diagram below shows a nephron from the kidney. What are the structures labelled X and Y? **[1]**

NOTES

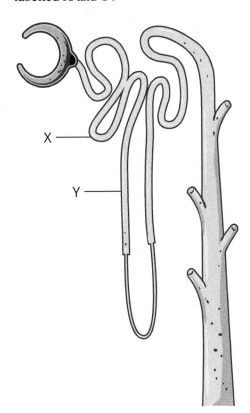

		X	**Y**
☐	A.	Glomerulus	Glomerulus
☐	B.	Proximal convoluted tubule	Loop of Henle
☐	C.	Loop of Henle	Collecting duct
☐	D.	Collecting duct	Proximal convoluted tubule

31. Which row below shows the correct excretory product paired with the correct animal? **[1]**

		Organisms	**Excretory product**
☐	A.	Fish	Uric acid
☐	B.	Mammals	Urea
☐	C.	Insects	Ammonia
☐	D.	Birds and reptiles	Urine

32. Which is the correct order for an electrical impulse in the heart? **[1]**

☐ A. Sinoatrial node → atrioventricular node → bundle of His → Purkinje fibres

☐ B. Atrioventricular node → sinoatrial node → bundle of His → Purkinje fibres

☐ C. Sinoatrial node → atrioventricular node → Purkinje fibres → bundle of His

☐ D. Atrioventricular node → sinoatrial node → Purkinje fibres → bundle of His

33. The image below is of a *Helianthus* stem. Identify tissues X and Y along with their function. **[1]**

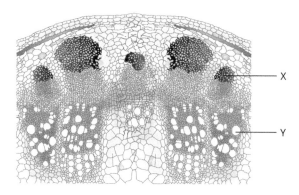

		X	**Y**
☐	A.	Phloem to transport water	Xylem to transport sugar
☐	B.	Xylem to transport sugar	Phloem to transport water
☐	C.	Phloem to transport sugar	Xylem to transport water
☐	D.	Xylem to transport water	Phloem to transport sugar

34. Which controls flowering in long-day plants? **[1]**

- ☐ A. P_r is converted by red light into P_{fr} promoting transcription of genes needed for flowering
- ☐ B. P_r is converted by red light into P_{fr} inhibiting transcription of genes needed for flowering
- ☐ C. P_{fr} is converted by red light into P_r promoting transcription of genes needed for flowering
- ☐ D. P_{fr} is converted by red light into P_r inhibiting transcription of genes needed for flowering

35. What is the correct definition of a community? **[1]**

- ☐ A. A group of organisms of the same species, living in the same geographical area at the same time
- ☐ B. A group of populations living together and interacting with each other within a given area at the same time
- ☐ C. The environment in which a species lives
- ☐ D. A group of populations interacting with each other and their abiotic environment

36. In the graph below, the X-axis represents the phenotypic trait and the Y-axis represents the number of organisms. What type of natural selection does this represent?

The original population is favouring extreme phenotypes over the intermediate. **[1]**

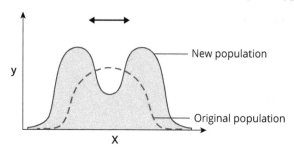

- ☐ A. Directional natural selection
- ☐ B. Stabilizing natural selection
- ☐ C. Disruptive natural selection
- ☐ D. Balancing natural selection

NOTES

37. Which of the following is the insoluble fibrous protein involved in blood clotting? **[1]**

- ☐ A. Prothrombin
- ☐ B. Thrombin
- ☐ C. Fibrinogen
- ☐ D. Fibrin

38. The diagram below shows a human egg.

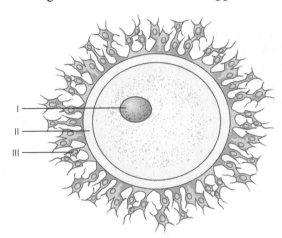

What are the structures labelled I, II and III? **[1]**

	I	II	III
☐ A.	Diploid nucleus	Follicle cells	Zona pellucida
☐ B.	Diploid nucleus	Zona pellucida	Follicle cells
☐ C.	Haploid nucleus	Follicle cells	Zona pellucida
☐ D.	Haploid nucleus	Zona pellucida	Follicle cells

39. The micrograph shows a seminiferous tubule. Where are the spermatozoa located? **[1]**

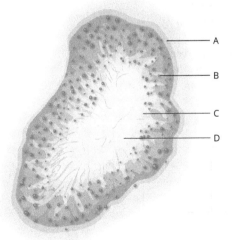

- ☐ A.
- ☐ B.
- ☐ B.
- ☐ D.

40. Which of these statements about the placenta is incorrect? **[1]**

- ☐ A. The placenta allows the transfer of active immunity from mother to fetus
- ☐ B. The placenta allows the transfer of glucose and oxygen from the blood from mother to fetus
- ☐ C. The placenta develops after the blastocyst has implanted
- ☐ D. The placenta produces the hormone oestrogen

Paper 2: Higher Level

- Set your timer for **2 hours and 15 minutes**
- The maximum mark for this examination paper is **[72 marks]**
- **Section A:** answer ALL the questions
- **Section B:** answer two questions
- A calculator is required for this paper

Section A

NOTES

1. Coronaviruses are a family of viruses that cause a range of illnesses from the common cold to Middle East Respiratory syndrome (MERS-CoV) and Severe Acute Respiratory Syndrome (SARS-CoV). The image shows the coronavirus (SARS-CoV-2) first reported from Wuhan, China, on 31 December 2019. This virus causes an illness called COVID-19 which can affect the respiratory system.

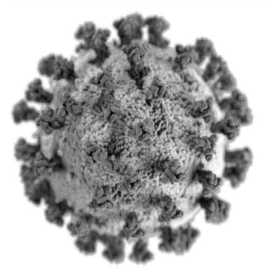

The table shows three coronaviruses that are zoonotic, meaning they are transmitted between animals and people. Detailed investigations found that SARS-CoV was transmitted from civet cats to humans and MERS-CoV from dromedary camels to human. Both bats and pangolins have been suspects for the outbreak of CoV-19 but have not been confirmed as the cause (shown by an *).

	Date	Countries affected	Origin	Cases	Deaths	Percentage deaths (%)
SARS-CoV	Nov 2002– July 2003	26	Bats to civet-cats-human	8,097	774	9.56
MERS-CoV	2012	27	Camels-humans	2494	858	34.40
CoV-19	2020	117	Bats to pangolin to human possibility	125,048*	4,613*	

[Source: WHO. Figures from 12 March 2020]

(a) Calculate the percentage of cases of CoV-19 that result in death. **[1]**

....

(b) Suggest why all three outbreaks are classified as pandemics. **[1]**

....

....

(c) Based on the data in the table, compare and contrast the severity of the three pandemics. **[3]**

....

....

....

....

....

The graphs show the epidemiological data for CoV-19 in China and Italy from 21 January 2020 to 11 March 2020.

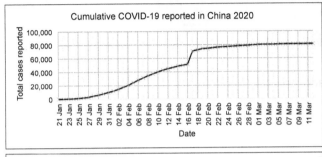

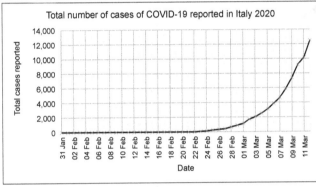

[Source: WHO. Data from 12 March 2020]

(d) State the number of cases in Italy on 3 March 2020 *and* 11 March 2020. **[1]**

....

....

(e) Calculate the percentage increase in the number of cases in Italy between 3 March and 11 March. **[1]**

....

The graph below shows the number of cases reported each day.

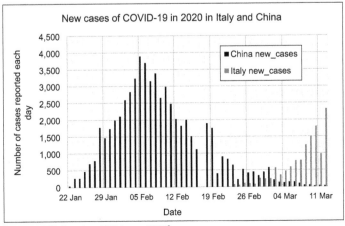

[Source: WHO, Data from 17 February 2020]

(f) Suggest one reason for the change in number of reported new cases in China after 5 February 2020. **[1]**

....

....

The table summarizes the epidemiological data from Italy and China on 5 February and 12 March 2020.

COVID-19	Country	Total cases	Total deaths	Percentage of deaths (%)
5 February 2020	China	24,363	491	2.0
	Italy	2	0	0
12 March 2020	China	80,981	3173	3.9
	Italy	12,462	827	6.6

[Source: WHO, https://www.who.int/docs/default-source/coronaviruse/20200312-sitrep-52-covid-19.pdf?sfvrsn=e2bfc9c0_2 https://www.who.int/docs/default-source/coronaviruse/situation-reports/20200205-sitrep-16-ncov.pdf?sfvrsn=23af287f_4]

(g) Suggest a reason why the death rate may not be accurate. **[1]**

....

....

(h) Based on all the data, compare and contrast the progress of the epidemic in the two countries. **[3]**

....

....

....

....

....

....

(i) The genome of the coronavirus COVID-19 was published on a global database by the Shanghai Public Health Clinical Center & School of Public Health on Gen bank.

State an advantage of publishing the genome on a global database. **[1]**

....

....

(j) Suggest a possible preventive measure for COVID-19. **[1]**

....

....

NOTES

2. (a) The image shows a cell undergoing meiosis. Identify the stage of meiosis it is undergoing. [1]

....

(b) Explain how meiosis can result in Down syndrome. [2]

....

....

....

....

3. (a) Label the parts of the heart indicated by I, II, III, IV. [2]

I ....

II ....

III ....

IV ....

(b) Distinguish between the structure of arteries and veins. [3]

....

....

....

....

....

....

(c) State the name of the artery that supplies the heart muscle with blood. [1]

....

(d) In coronary heart disease (CHD) a cholesterol plaque in the artery that supplies the heart could burst resulting in a blood clot forming at the site of the rupture.

Outline the process of blood clotting. [3]

....

....

....

....

....

NOTES

4. (a) Obesity is a risk factor for CHD and can be caused by overeating.
Explain how appetite is controlled. **[3]**

(b) Define excretion. **[1]**

(c) Distinguish between the nitrogenous waste of freshwater fish and
mammals. **[2]**

5. The light micrograph below is of a *Helianthus*.

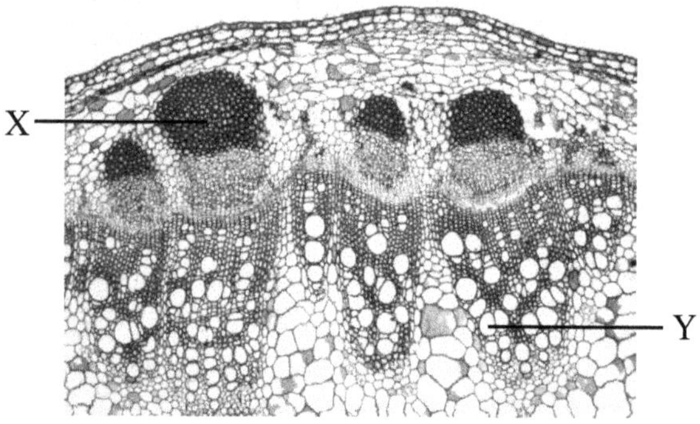

(a) Determine the organ this is from. **[1]**

(b) Identify tissue X and tissue Y and state what each tissue transports. **[2]**

(c) State one difference between the structure of the two tissues. **[1]**

(d) State the type of bond that occurs between water molecules. **[1]**

(e) Outline the importance of cohesion and adhesion in transpiration. **[3]**

NOTES

Section B

Answer **two** questions from a choice of three. Up to one additional mark is available for the construction of your answers for each question.

6. Enzymes are proteins that are expressed in certain cells and not in others.

   (a) Distinguish between globular and fibrous proteins. [3]

   (b) Outline end product inhibition in a metabolic pathway. [4]

   (c) Explain the regulation of gene expression in eukaryotic cells. [8]

7. Pneumonia is a bacterial disease that affects gas exchange in the lungs.

(a) Distinguish between type I and type II pneumocytes. [3]

....

....

....

....

....

....

(b) Outline the mechanism of ventilation. [5]

....

....

....

....

....

....

....

....

....

....

(c) Describe how antibiotic-resistant bacteria give evidence for evolution. [7]

....

....

....

....

....

....

....

....

....

....

....

....

....

NOTES

8. Some crops have been genetically modified to be resistant to pesticides.

(a) Draw the structure of a sarcomere. **[4]**

(b) Explain the mechanism of contraction of skeletal muscle. **[6]**

....

....

....

....

....

....

....

....

....

....

....

....

(c) Outline the use and environmental impact of neonicotinoid pesticides. **[5]**

....

....

....

....

....

....

....

....

....

....

Paper 3: Higher Level

- Set your timer for **1 hour and 15 minutes**
- The maximum mark for this examination paper is **[45 marks]**
- **Section A:** answer ALL the questions
- **Section B:** answer all of the questions from ONE of the options
- A calculator is required for this paper

Section A

NOTES

1. Thin-layer chromatography was used to analyse photosynthetic pigments including carotene, xanthophylls and chlorophylls. Different seaweeds were tested including green seaweed, red seaweed and brown seaweed.

   Key: Green seaweed *Ulva pertusa (Upe)*, red seaweed *Bangia fuscopurpurea (Bp)* and *Porphyra yezoensis (Py)*, and brown seaweeds *Ectocarpus siliculosus (Es)* and *Undaria pinnatifida (Upi)*.

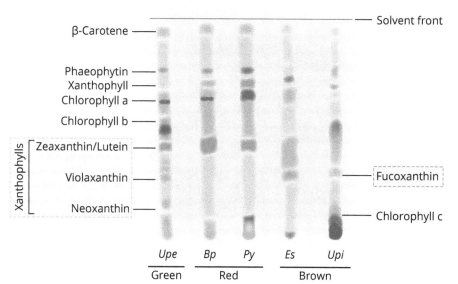

[Source: Mikami, K,, Hosokawa, M., 'Biosynthetic pathway and health benefits of fucoxanthin, an algae-specific xanthophyll in brown seaweeds'. Int J Mol Sci. 2013;14(7):13763-13781. Published 2013 Jul 2. doi:10.3390/ijms140713763, (CC BY 3.0)]

(a) Calculate the R_f value for chlorophyll a in *Ulva pertusa*. **[1]**

(b) Suggest a suitable solvent. **[1]**

(c) Identify a pigment that is only present in the brown seaweeds. **[1]**

Seaweeds submersed in water need additional pigments such as Fucoxanthin to trap light energy as the surface of water reflects away a lot of the blue light, and red light does not penetrate far into the water. Green light penetrates the furthest. The absorption spectrum for chlorophyll a, chlorophyll b and Fucoxanthin is shown below.

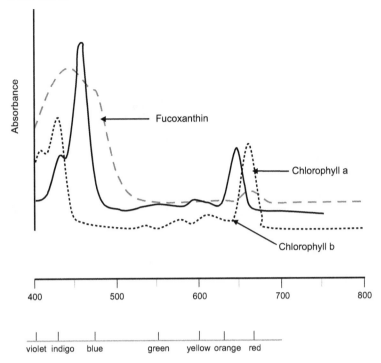

(d) Compare and contrast the spectrum for chlorophylls a and b.   **[2]**

....

....

....

....

....

....

(e) The brown algae at low shore are covered by large amounts of water at high tide. Suggest an advantage Fucoxanthin gives some seaweeds.   **[1]**

....

....

....

2. The equipment below was used to investigate the effect of wind on the transpiration rate of a tomato plant.

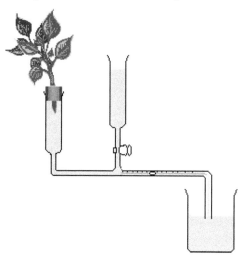

(a) State the name of the equipment. **[1]**

....

(b) State a variable that the student should control. **[1]**

....

(c) Suggest why the student left the plant for 5 minutes in each condition before recording any results. **[1]**

....

....

The graph below shows the results of the experiment.

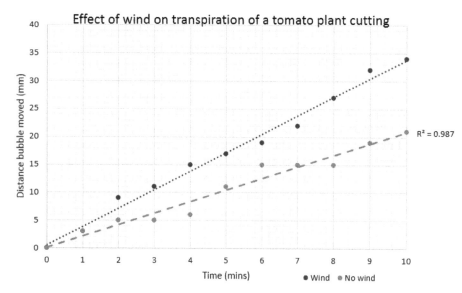

(d) Calculate the rate of transpiration in the wind condition in mm/min. **[1]**

....

(e) This method is an indirect method of measuring transpiration. Suggest a limitation of using an indirect method. **[1]**

....

....

NOTES

The image below shows a scanning electron micrograph of a stoma in a tomato leaf

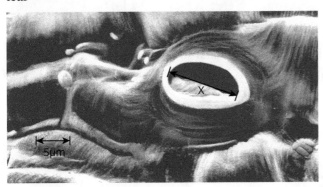

(f) Calculate the diameter of the stomatal pore where indicated by X.   **[1]**

....

3. In 1952, Stanley Miller and Harold Urey gathered evidence about the chemical origin of life. They modelled the early atmosphere using water, methane, ammonia and hydrogen. High temperatures and electricity were also used to simulate the climate on early Earth.

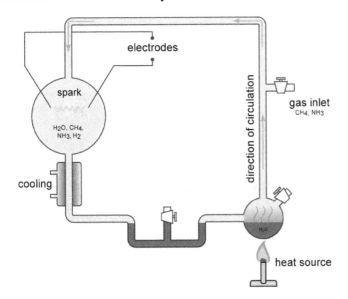

(a) State the name of the molecule Miller and Urey synthesized.   **[1]**

....

(b) Outline the importance of their discovery.   **[1]**

....

....

(c) Suggest a limitation of this model.   **[1]**

....

....

NOTES

Section B

Option A – Neurobiology and behaviour

1. Approximately 1 in 4,000 babies has a perinatal stroke shortly after or during birth. A study investigated teenagers and young adults who had a perinatal stroke.

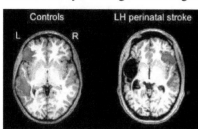

Images of the brain were taken when given a language task.

[Source: Elissa Newport/The Pediatric Stroke Research Project]

(a) Identify the name and location of the area of the brain that is present in the control subject and missing in the stroke subject. [1]

....

(b) State the name of the scan used. [1]

....

(c) Determine if the person who had the perinatal stroke is able to convert thoughts to speech. [2]

....

....

....

2. The image below is a diagram of the retina.

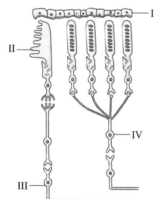

(a) Identify structures I, II, III and IV. [2]

I ....   II ....

III ....   IV ....

(b) Explain the function of bipolar cells and ganglion cells in response to light and dark. [4]

....

....

....

....

....

....

....

....

3. Coding of auditory stimuli by neurons in the medial geniculate body (MGB) of the auditory cortex is dependent on the neurotransmission of GABA. A study investigated the expression of the neurotransmitter GABA in the auditory cortex of young, Y (3–8 months), and aged, A (28–32 months), rats.

The graph shows the relative protein density of the enzyme GAD_{67}. GAD_{67} is found in the presynaptic knob and is important for the production of GABA for neuron activity.

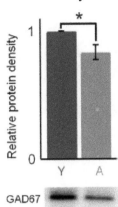

[Source: Adapted from Figure 1: Richardson, Ben and Ling, et al. (2013). 'Reduced GABA(A) Receptor-Mediated Tonic Inhibition in Aged Rat Auditory Thalamus'. *The Journal of neuroscience : the official journal of the Society for Neuroscience.* 33. 1218-27. 10.1523/JNEUROSCI.3277-12.2013.]

(a) Deduce the effect of aging on the production of the neurotransmitter GABA. **[1]**

....

....

Below is the neurotransmission trace showing the postsynaptic potentials caused by the neurotransmitter GABA in young and aged rats.

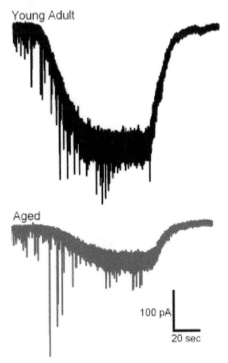

[Source: Adapted from Figure 5: Richardson, Ben and Ling, et al. (2013). 'Reduced GABA(A) Receptor-Mediated Tonic Inhibition in Aged Rat Auditory Thalamus'. *The Journal of neuroscience : the official journal of the Society for Neuroscience.* 33. 1218-27. 10.1523/JNEUROSCI.3277-12.2013.]

(b) Suggest, with a reason, the type of postsynaptic potential caused by GABA. **[1]**

....

....

Graphs A and B compare the resting membrane potential of the postsynaptic neurons in the medial geniculate body (MGB) which is responsible for the processing and filtering of acoustic information.

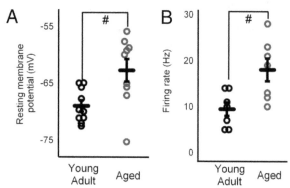

[Source: Adapted from Figure 5: Richardson, Ben and Ling, et al. (2013). 'Reduced GABA(A) Receptor-Mediated Tonic Inhibition in Aged Rat Auditory Thalamus'. *The Journal of neuroscience : the official journal of the Society for Neuroscience*. 33. 1218-27. 10.1523/JNEUROSCI.3277-12.2013.]

(c) Outline the effect of aging on the postsynaptic neurons. **[2]**

....

....

....

....

(d) Suggest the mechanism for the effects caused by changes in the expression of GABA in the MGB aged rats. **[3]**

....

....

....

....

....

....

4. Oystercatchers (*Haematopus ostralegus*) forage on mussels (*Mytilus edulis*) and crack them open with their bills. The constraints on these birds are the characteristics of the different mussel sizes. Large mussels give more energy than small mussels but large mussels have thicker shells. A study investigated the size of muscles foraged and the time it took to open the shell.

(a) State the type of behaviour the oystercatchers are showing. **[1]**

....

....

NOTES

The graph below shows the relationship between mussel length and shell thickness of the mussels present in the plot and the mussels opened by oystercatchers. The open circles represent mussels opened by oystercatchers, the closed represent mussels present in the area. Error bars are ± 1 standard error from the mean.

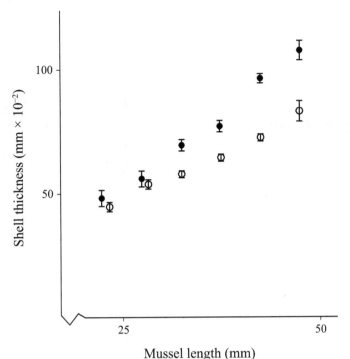

Shell thickness (mm × 10⁻²) vs Mussel length (mm)

[Source: Meire, P. and A. Ervynck. "Are oystercatchers (Haematopus ostralegus) selecting the most profitable mussels (Mytilus edulis)?" Animal Behaviour 34 (1986): 1427-1435.]

(b) Outline the trend between mussel shell length and thickness.   **[1]**

....

....

(c) Outline how the thickness of the mussel shell affects the proportion of mussels opened.   **[2]**

....

....

....

....

The profitability of food is measured as the mg of food per time taken to obtain the food.

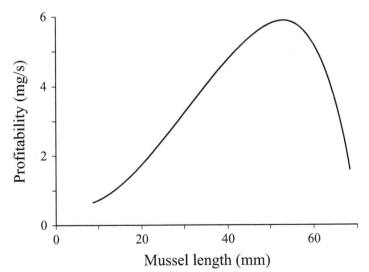

Profitability (mg/s) vs Mussel length (mm)

(d) Determine, with reasons, the optimum mussel shell length suggested by this graph. [3]

....

....

....

....

....

(e) Explain how sound waves are perceived by the ear and transmitted to the brain. [6]

....

....

....

....

....

....

....

....

....

....

....

NOTES

Option B – Biotechnology and bioinformatics

1. The causes of obesity are frequently researched as obesity is associated with many health disorders. The animal model ob/ob mice exhibit a selectively bred recessive mutation in the leptin gene, meaning leptin is unable to bind to its receptors. Leptin is a hormone that controls appetite.

   The ob/ob mouse is on the left.

   (a) Deduce the effect of leptin deficiency on obesity. **[1]**

   ....

   ....

   (b) Evaluate the advantages and disadvantages of using of using the ob/ob model of mice in the study on leptin. **[5]**

   ....

   ....

   ....

   ....

   ....

   ....

   ....

   ....

   ....

   (c) Outline how knockout mice are useful in medical research. **[3]**

   ....

   ....

   ....

   ....

   ....

   ....

NOTES

2. The image shows the treatment of a child with Severe Combined Immunodeficiency Disorder (SCID) in 1971. SCID is an inherited disorder resulting from the lack of an enzyme adenosine deaminase (ADA). The lack of ADA enzyme results in the build-up of a toxin (deoxyadenosine), which results in low numbers of functional T and B lymphocytes.

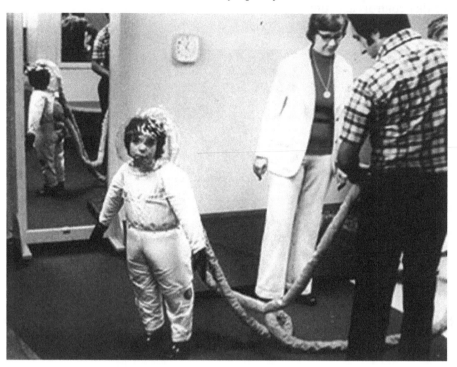

(a) Outline the use of viral vectors in gene therapy to treat SCID. **[4]**

....

....

....

....

....

....

....

(b) Discuss the advantages and disadvantages of using viral vectors. **[3]**

....

....

....

....

....

(c) Outline how expressed sequence tags (ESTs) can aid gene research. **[2]**

....

....

....

3. The evolutionary trees below are based on a BLAST search for a nucleotide sequence for the protein actin among some species. The same tree is shown with binomial names and common names.

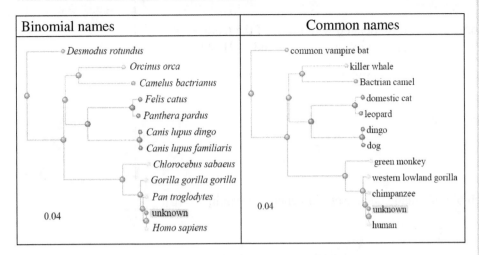

Binomial names	Common names

(a) Suggest why the binomial name is used in phylogenetics. **[1]**

....

....

(b) Determine, with a reason, whether the tree is a cladogram or a phylogram. **[1]**

....

....

(c) State, with a reason, which species are more closely related to each other, *Orcinus orca* and *Camelus bactrianus* or *Felis catus* and *Panthera pardus*. **[1]**

....

....

4. Below is an ideogram of chromosome 21.

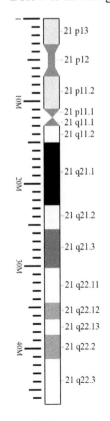

(a) State the approximate number of base pairs on chromosome 21 from this ideogram. [1]

....

The area of chromosome 21 that is associated with many of the features of Down syndrome is known as the critical region and is found at 21q22.3.

(b) Draw an arrow on the diagram on the previous page to the band pointing to the critical region. [1]

(c) Determine what p and q represent. [1]

....

....

(d) Outline ways in which bioinformatic databases aid genetic research. [6]

....

....

....

....

....

....

....

....

....

....

....

....

....

....

....

....

....

....

....

....

....

....

....

NOTES

Option C – Ecology and conservation

1. The grey squirrel (left image) *Sciurus carolinensis* is native to North America and was introduced to the United Kingdom in 1876. They eat seeds, flowers, fruit, fungi and insects. They can live in any species of tree and also carry a poxvirus. The smaller red squirrel *Sciurus vulgaris* is native to the United Kingdom. It lives in conifer trees and eats seeds, flowers, fruit and fungi.

Below are maps showing the distribution of the red squirrels and grey squirrels in 1945 and 2010.

 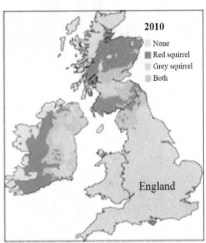

[Source: Maps showing grey squirrel spread and red squirrel decline 1945 & 2010 by Craig Shuttleworth & Red Squirrel Survival Trust]

(a) State the name given to a non-native invasive species that has been introduced to an area where it becomes established. **[1]**

....

(b) Use the graph to outline the competition between the red squirrel and grey squirrel between 1945 and 2010 in England, Scotland, Wales and Ireland. **[2]**

....

....

....

....

(c) Suggest why the decline in red squirrels is lower in some areas. **[1]**

....

....

The European pine marten (*Martes martes*) is a native predator of red squirrels but eats both red and grey squirrels. A study in 2015 at twenty sites in Northern Ireland investigated the number of visits of both red and grey squirrels to feeders before and after pine marten scent had been added to the feeding station. In total 6,076 visits were recorded by grey squirrels and 4,363 by red squirrels.

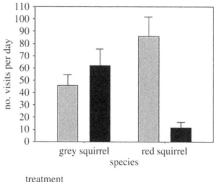

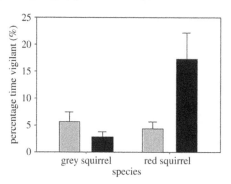

treatment
□ pre-treatment
■ post-treatment

[Source: Twining, Joshua P., Ian, Montgomery W.., Price, Lily, Kunc, Hansjoerg P. and Tosh, David G., 2020, 'Native and invasive squirrels show different behavioural responses to scent of a shared native predator', R. Soc. open sci.7191841191841, http://doi.org/10.1098/rsos.191841]

(d) Outline the difference in reaction to the pine marten scent in both squirrel species. **[1]**

(e) State the name given to an organism introduced to an area to reduce numbers. **[1]**

(f) Evaluate whether the re-introduction of pine martens to England would be useful for controlling populations of grey squirrels. **[2]**

2. A simplified food web for North American lakes is outlined below.

Microscopic photosynthetic plants (phytoplankton) and algae trap light energy for photosynthesis. Microscopic animals (zooplankton) feed on the phytoplankton and the algae. Shrimp feed on the zooplankton. Zooplankton are a valuable food source for planktivorous fish such as the piscivorous fish.

(a) State the mode of nutrition of phytoplankton. **[1]**

(b) Draw a food web for the organisms stated in North American lakes. **[2]**

(c) Define the term limiting factor. **[1]**

(d) Identify a top-down and bottom-up limiting factor on phytoplankton. **[1]**

(e) Suggest a human activity that increases the level of nitrates or phosphates in the water. **[1]**

(f) Outline eutrophication of the lake water. **[4]**

3. The image shows part of a food web from a rocky shore.

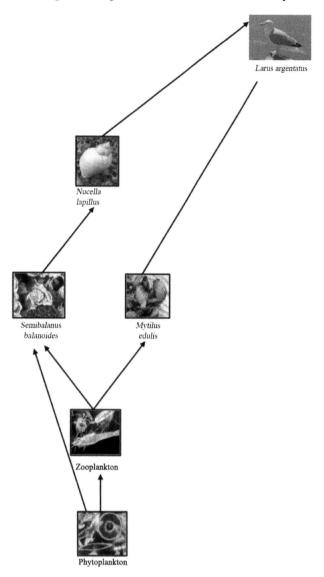

(a) State the possible trophic levels of dogwhelks (*Nucella lapillus*). **[1]**

....

....

(b) Identify a top-down and bottom-up limiting factor on *Nucella lapillus*. **[1]**

Top-down: ....

Bottom-up: ....

(c) Outline a method to estimate the population size of the motile species
Nucella lapillus. **[4]**

....

....

....

....

....

....

....

....

4. Draw and label the nitrogen cycle. [6]

Option D – Human physiology

1. A study investigated how the level of cholesterol in the blood affected death rate from coronary heart disease (CHD) in middle aged men in different populations.

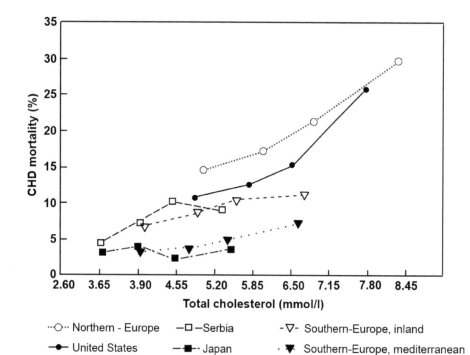

[Source: Seven Countries Study]

(a) State the overall trend between cholesterol and CHD mortality. **[1]**

....

....

(b) Identify the area whose population is most at risk of CHD. **[1]**

....

(c) Compare and contrast the data for Southern Europe, inland and Southern Europe, mediterranean. **[2]**

....

....

....

....

(d) Suggest a reason for the differences in data in Southern Europe, inland and Southern Europe, mediterranean. **[1]**

....

....

(e) Evaluate this hypothesis: 'Cholesterol causes CHD'. **[2]**

....

....

....

....

2. The image below shows the electrical activity in different regions of the heart during a normal heart cycle.

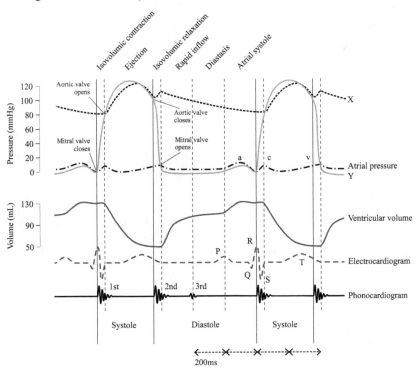

(a) Identify what is represented by X and Y. [2]

X: ....

Y: ....

(b) Determine the time delay between the P and the QRS complex. [1]

....

(c) Outline the need for the delay between the P and the QRS complex. [1]

....

....

(d) Explain how valves control the flow of blood through the heart. [2]

....

....

....

....

3. The graph shows the oxygen dissociation curve for hemoglobin (Percentage of oxy-hemoglobin %) as the partial pressure of oxygen in the surrounding tissues changes.

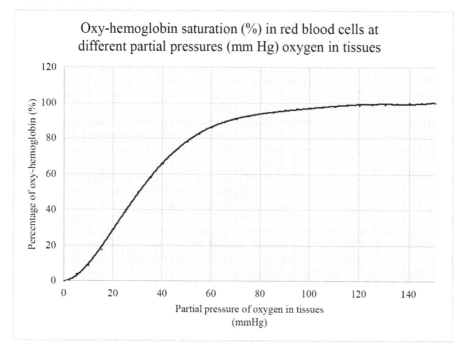

Oxy-hemoglobin saturation (%) in red blood cells at different partial pressures (mm Hg) oxygen in tissues

(a) The alveoli have a partial pressure of 104 mmHg. State the percentage of hemoglobin that is oxygenated in the lungs. **[1]**

(b) Explain the movement of oxygen between the alveoli and capillaries in the lungs if the blood in the capillaries has an initial partial pressure of 40 mmHg. **[2]**

The graph shows the oxy-hemoglobin saturation in red blood cells at different partial pressures.

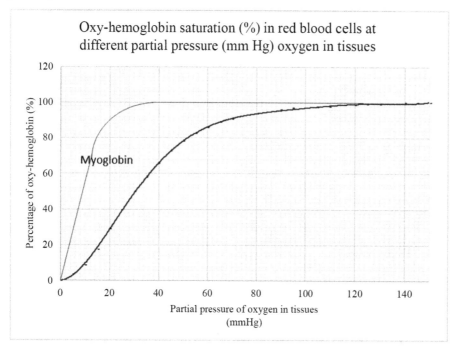

Oxy-hemoglobin saturation (%) in red blood cells at different partial pressure (mm Hg) oxygen in tissues

(c) Explain the significance of the hemoglobin and myoglobin oxygen
dissociation curves in exercising muscle tissue. [4]

....

....

....

....

....

....

....

4. (a) State the name of the ductless glands that release hormones directly
into the bloodstream. [1]

....

(b) Compare and contrast the mechanism of action of peptide and steroid
hormones. [3]

....

....

....

....

....

....

5. Normal blood pH is pH 7.4 (range 7.35–7.45). Describe the mechanisms
involved in the regulation of the pH of the blood during intense exercise. [6]

....

....

....

....

....

....

....

....

....

....

NOTES

Answers

Set A

Set A: Paper 1

Question no.	Answer	Question no.	Answer	Question no.	Answer
1.	C	15.	C	29.	B
2.	B	16.	A	30.	B
3.	A	17.	D	31.	B
4.	B	18.	B	32.	D
5.	D	19.	C	33.	C
6.	A	20.	B	34.	D
7.	B	21.	C	35.	A
8.	B	22.	A	36.	C
9.	B	23.	D	37.	C
10.	C	24.	C	38.	C
11.	C	25.	A	39.	C
12.	B	26.	D	40.	C
13.	D	27.	B		
14.	B	28.	A		

Set A: Paper 2

Section A

1. (a) Eukaryota **[1]**
   (b) Antibiotics **[1]**
   (c) May 2005 **[Both month and year needed for 1]**
   (d) (i) 4 ÷ 27 = 14.8% **[1]**
   (d) (ii) 1 ÷ 31 = 3.23% **[1]**
   (e) Similarities:
   - No TB found in either group <u>Jan 05–Mar 05/or in first two months</u>
   - Both had increases in TB near the end of the study
   - Both had no cases of TB in <u>winter/(Jan–Mar)</u>
   - Both had more possums not infected than infected
   Differences:
   - The first cases of TB are apparent in May in the control group, whereas TB is found in the vaccinated group in September
   - Higher number of cases of TB are found in the control group and it is the converse for the vaccinated group **[Max 2]**
   (f) 600 pg/ml **[1]**
   (g) • Antibodies are higher in the vaccinated calves (peak 1250 pg/ml) compared to non-vaccinated (peak 600 pg/ml)
   - It is 106 weeks before antibody levels match the non-vaccinated calves/vaccine lasts long enough for practical reapplication **[2]**
   (h) • Effective in vaccinated calves as more <u>interferon gamma/antibody</u> is made
   - Effective as the antibody/interferon gamma is made <u>faster</u>
   - Not effective as offers short-term immunity/in both the non-vaccinated and BCG vaccinated the immune response/antibody production/IFN-γ drops significantly by 25 weeks
   - Not effective as no significant difference between non-vaccinated/control and vaccinated at 50+ weeks **[Max 3]**

2. (a) Urea, vitalism **[Both needed for 1 mark]**
   (b) • Osmolarity is the number of osmoles of a solute per litre of solution / concentration of osmotically active solutes per litre / the number of moles of solute that contribute to the osmotic pressure of a solution
   - osmol/L or mosmol/L or Osm/L or mOsm/L **[Max 2]**
   (c) • Osmoconformers internal solute conditions/osmolarity changes whereas osmoregulators keep their internal solute concentration/osmolarity constant
   - Osmoregulators can have different osmolarity to the environment whereas osmoconformers cannot
   - Osmoconformers include mussels, scallops, crabs, jellyfish, squid, sharks, skates, hagfish, starfish
   - Osmoregulators include mammals, birds, freshwater fish **[Max 2]**

3. (a) • <u>DNA</u> wrapped around <u>histone</u> protein
   - 8 histone proteins/octomer made of 4 different types of protein
   - Linker DNA connecting nucleosomes
   - Additional H1 histone
   - N-terminal tails extrude out of protein

   - Accept labelled diagram **[Max 2]**

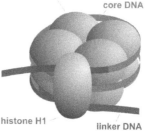
octamer of core histones:
H2A, H2B, H3, H4 (each one ×2)
core DNA
histone H1   linker DNA

   (b) • Methylation decreases (gene expression)/switches off genes whereas acetylation increases (gene expression)/switches on genes
   - Genes are turned on when they are acetylated and unmethylated / genes are turned off when they are deacetylated and methylated
   - Methylation involves adding a methyl/CH$_3$ group, acetylation involves adding an acetyl/CH$_3$C=O group
   - Methylation binds to DNA /cytosine, acetylation binds to histone protein
   - Methylation causes heterochromatin/supercoiled DNA whereas acetylation causes euchromatin/ loosely packed / not supercoiled DNA **[Max 3]**
   (c) • Methylation inhibits transcription/expression of tumour suppressor genes
   - So proteins that inhibit cell division/mitosis are not made
   - So uncontrolled cell division/mitosis occurs **[2]**
   (d) 10/25 × 100 = 40% **[1]**

4. (a) Transmission <u>electron</u> microscope as virus is too small to be seen with a light microscope / higher resolution and magnification than light microscope. **[1]**
   (b) • Vaccine is a weak/attenuated form of the virus
   - Antigen on membrane of pathogen/antigen is displayed on membrane of a phagocyte/antigen-presenting cell
   - T cells/T$_H$ cells/T lymphocytes bind to the antigen / T cells/T$_H$ cells/T lymphocytes release cytokines/activate specific B cells/B lymphocytes
   - B cells/B lymphocytes divide by mitosis/produce clones
   - Forming plasma cells/B cells that release antibodies
   - Antibody binds to virus antigens causing destruction of virus/agglutination/cell lysis/precipitation/perforation/neutralization
   - Some B cells make memory cells
   - Memory cells make antibodies faster / more antibodies upon exposure to coronavirus **[Max 5]**
   (c) • Antibiotics inhibit metabolic pathways/metabolism in cells
   - Viruses don't have their own metabolic pathways/metabolism / Viruses use the host cells' metabolic pathways/metabolism
   - Antibiotics do not inhibit/block host cell metabolism **[Max 2]**

5. (a) I – ovary
   II – style
   III – anther **[All three correct for 1 mark]**
   (b) Angiospermophytes/angiospermophyta **[1]**
   (c) • Insect-pollinated as colourful petals to attract insects
   - Insect-pollinated as pollen on anthers inside flower so will brush onto insects (as they make way to nectary)
   - Insect-pollinated as stigma inside flower so will brush onto insects (as they make way to nectary) **[Max 2]**
   (d) • The first name represents the genus and the second the species
   - Unique combination of two names designates species worldwide/internationally
   - They belong to the same genus/genus *Hyacinthoides*
   - They are different species **[Max 2]**

Section B

6. (a) Diagram could include:
   - Phosphate
   - Deoxyribose sugar
   - Nitrogenous base / adenine, thymine, cytosine, guanine

DNA nucleotide

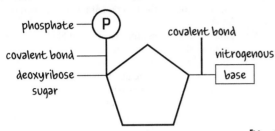

[Max 3]

(b)
- Enzyme DNA gyrase reduces tension in DNA
- Enzyme <u>DNA helicase</u> unwinds DNA / Enzyme <u>DNA helicase</u> breaks hydrogen bonds
- <u>DNA primase adds primers</u> to the lagging strand
- <u>DNA polymerase III</u> adds nucleotides in a 5 carbon to 3 carbon direction / joins the sugar–phosphate backbone/ phosphodiester bond / covalent bonds between sugars of one nucleotide and phosphate of another
- <u>DNA polymerase</u> I removes the primers
- <u>DNA ligase joins the Okazaki fragments</u>   **[Max 4]**

(c)
- Genetic modification involves taking the gene from one species and placing it in another species
- Plasmid removed from bacteria/*E.coli*/*Escherichia coli*...
- ...with restriction enzyme/endonuclease
- Human gene for insulin isolated / or other medical example
- The use of the same restriction enzyme/endonuclease results in <u>sticky ends</u> on the plasmid
- The gene is inserted into the plasmid / the sticky ends join by <u>complementary base pairing</u>
- <u>DNA ligase</u> joins the sugar–phosphate backbone/ phosphodiester bond
- <u>Transformed/recombinant</u> plasmid inserted into bacteria/*E. coli*
- Genetically modified bacteria are incubated and divide by binary fission
- Genetically modified bacteria secrete insulin which is purified
- Insulin is a <u>hormone</u> used/injected to treat <u>Type I diabetes</u>
- Insulin causes liver cells and muscle cells to take up glucose and convert it to <u>glycogen</u> for storage
- Insulin results in lowered blood sugar
  [Maximum if no reference to a disease is 5] [Max 8]

7. (a)
- Carbon dioxide
- Urea
- Hormones
- Antibodies/immunoglobulins
- Urea
- Water
- Glucose/amino acid   **[Max 3]**

(b)
- Damaged tissue/platelets release clotting factors
- Platelets form a clot over the damaged tissue
- Clotting factors start a cascade of reactions
- Soluble prothrombin is converted into the enzyme thrombin
- Thrombin catalyses the conversion of inactive/soluble fibrinogen into active/insoluble fibrin
- Fibrin forms a mesh/network of fibres over the damaged area/wound/clot
- Fibrin traps red blood cells and white blood cells, forming a scab
- Blood clot prevents the entry of pathogens / acts as a physical barrier   **[Max 5]**

(c)
- Inherited <u>recessive sex-linked</u> condition
- Gene for hemophilia is on X chromosome
- Results in lack of clotting factor VIII / blood doesn't clot
- X^H = allele for normal blood clotting
- X^h = hemophiliac allele
- Y = male
- Only females can be carriers / females have two copies of the X chromosome so can mask faulty allele
- More males are hemophiliac than females
- Males have one copy of X chromosome so cannot mask faulty allele / Males only need one copy of the gene to have hemophilia
- Males cannot pass on hemophilia gene to sons / males pass Y chromosomes to sons
- Sons inherit hemophilia from mothers only

Phenotype	Genotype
Normal female	$X^H X^H$
Carrier female (normal phenotype)	$X^H X^h$
Hemophiliac female	$X^h X^h$
Normal male	$X^H Y$
Hemophiliac male	$X^h Y$

- Example of a cross identifying parent phenotypes, genotypes and offspring e.g. carrier female ($X^H X^h$) × normal male ($X^H Y$) results in 1 normal female ($X^H X^H$) : 1 carrier female ($X^H X^h$), 1 hemophiliac male ($X^h Y$) : 1 normal male ($X^H Y$). Punnett grid or other valid example can be used:

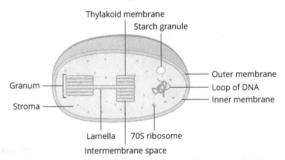

Parent genotype	$X^H X^h$		
	Gametes	X^H	X^h
$X^H Y$	X^H	$X^H X^H$	$X^H X^h$
	Y	$X^H Y$	$X^h Y$

[Max 7]

8. (a) Diagram could be/have:
- Oval shaped
- Inner and outer membrane labelled
- Intermembrane space labelled
- Stroma labelled
- Thylakoids/grana labelled
- Loop of DNA labelled
- 70S ribosome labelled

Chloroplast

Thylakoid membrane
Starch granule
Granum
Stroma
Outer membrane
Loop of DNA
Inner membrane
Lamella
70S ribosome
Intermembrane space

[Max 3] [Must be labelled]

(b)
- Occurs in thylakoid
- Photoactivation of Photosystem II / Light hits Photosystem II
- <u>Photolysis</u> involves splitting of water using light energy
- (H_2O is split) into 2 hydrogen ions/protons/$2H^+$ and oxygen atom/$O/^1/_2\,O_2$/ and <u>2</u> electrons/e^- /$2H_2O \rightarrow O_2 + 4H^+ + 4e^-$
- The electrons are excited/high energy
- Electrons are passed to the electron transport chain/electron carriers on the <u>thylakoid</u> membrane to <u>Photosystem I</u>
- The energy is used to pump protons/H^+/hydrogen ions into the thylakoid space from the stroma by active transport
- A proton gradient is built up / a high proton concentration occurs inside the thylakoid space
- Protons/H^+/hydrogen ions diffuse/move down the concentration gradient...
- ...through enzyme ATP synthase
- $ADP+P_i$ into ATP/photophosphorylation
- Production of ATP from ATP synthase is called chemiosmosis
- Light hits Photosystem I exciting low energy electrons
- The excited electrons from Photosystem I reduce NADP/form NADPH +H^+/ NADP +$2H^+$ + 2 $e^- \rightarrow$ NADPH + H^+
- The high energy electron, NADP and hydrogen ions/ protons/$2H^+$/ form reduced NADPH/NADPH+H^+
- ATP and NADPH pass to the light-independent stage of photosynthesis
- O_2 is a waste product   **[Max 7]**

(c)
- Flowers develop from <u>meristems</u> (at the apex) of the shoot
- Meristem cells then differentiate into flower tissue
- e.g. stamen and carpal and petals and sepal
- Changes in the length of nights/photoperiodism triggers <u>gene expression</u>
- Phytochromes exist in two (interchangeable) forms, P_{fr} and P_r
- P_{fr} absorbs red light/light with a wavelength of 660 nm
- Daylight has more red light/660 nm than far red/725 nm
- In daylight inactive P_{fr} is quickly converted to active P_{fr}
- At night active P_{fr} is converted back to inactive P_r

- In a short-day plant P$_{fr}$ inhibits flowering
- In a short-day plant a long night/night longer than a critical length is needed to reduce the amount of P$_{fr}$
- Example of a short-day plant is a chrysanthemum
- In a long-day plant P$_{fr}$ promotes flowering
- In a long-day plant a short night/day longer than a critical length is needed increase the amount of P$_{fr}$
- Example of a long-day plant is a carnation
- In far red light inactive P$_{fr}$ is quickly converted back to active P$_{fr}$ **[Max 5]**

Set A: Paper 3

Section A

1. (a) Scale bar image size = 16 mm
16 × 1,000 = 16,000 µm
Magnification = image size/actual size = 16,000/5 = × 3,200
Answer: × 3,200 **[1]**
(b) Actual size = image size/magnification
Image size = diameter = 27 mm × 1,000 = 27,000 µm
Actual size = 27,000/3,200 = 8.43 µm **[1]**
(c) • Manufacture lactose-free milk
- Increase the sweetness of yogurt/dairy-based sweets
- Improve the texture of ice-cream by reducing crystallization
- Speed up fermentation in yogurt/cheese **[Max 1]**

2. (a) Concentration of lactose / concentration of lactase **[1]**
(b) (i) • Both have the optimum pH of 7
- Both have similar/same relative enzyme activity at pH 6.5 and 7
- Immobilized lactase has a significantly higher enzyme activity at pH 5.5–6
- Immobilized lactase has a significantly higher enzyme activity at pH 7.5 and above **[Max 2]**
(ii) • The lactase can be reused
- Greater thermal and pH stability so less enzymes are denatured
- Easy to purify the product/separate the enzyme from the product **[Max 1]**

3. (a) I – ureter
II – pelvis
III – cortex
IV – medulla **[1 mark for every 2 correct answers, Max 2]**
(b) Renal artery **[1]**

4. (a) Scale bar is 10 mm = 10 000 µm
Actual size of scale bar is 200 nm = 0.2 µm
Magnification = image size/actual size
Magnification = 10 000/0.2
Magnification = x 50 000 **[1]**
(b) Transmission electron microscope as it has a high magnification/high resolution/to see internal detail. **[1]**
(c) To kill it / to stop photosynthesis / to stop the Calvin cycle / to stop metabolism. **[1]**
(d) • PGA
- The PGA is present in the largest quantity after the shorter exposure/ 5 seconds indicates that it is the first stable product
- At 30 seconds the other products appear after PGA so must be derived from it. **[2]**

Section B

Option A: Neurobiology and behaviour

1. (a) Neural tube **[1]**
(b) • Neurulation only occurs in chordates
- There are three germ layers – ectoderm, mesoderm and endoderm
- Neurulation is the formation of neural tube from the ectoderm
- Neurulation is controlled by the notochord / the notochord is part of the mesoderm
- Ectoderm cells differentiate to form a neural plate
- The neural plate folds inwards/makes a neural groove
- Ends of the fold/plate join together to form a neural tube
- The neural tube separates from the ectoderm
- The neural tube elongates
- The neural tube cells differentiate into brain and spinal cord/central nervous system
- The neural crest differentiates into peripheral nervous system **[Max 4]**

2. (a) Nucleus accumbens **[1]**
(b) Cerebellum **[1]**
(c) • Inhibitory synapses use GABA as their neurotransmitter
- Alcohol increases the function of inhibitory neurotransmitters/GABA
- GABA diffuses across the synapse and binds to GABA receptors on the postsynaptic membrane
- Opening voltage-gated chloride channels
- Chloride ions enter the postsynaptic neuron
- Causing it to be hyperpolarized/–80 mV / less likely to depolarize
- Alcohol binds to GABA receptor at allosteric site
- Increasing the opening of Cl⁻ channels hyperpolarizing the postsynaptic neuron
- Making it less likely to reach a threshold potential
- Dopamine and glutamate are excitatory neurotransmitters
- Alcohol also decreases excitatory neurotransmission of glutamate
- Alcohol is addictive as it increases neurotransmission of dopamine in the reward centre
- Dopamine stimulates/reward centre is artificially stimulated / making person feel they need the feelings of pleasure **[Max 4]**

3. (a) I – iris
II – cornea
III – lens
IV – optic nerve **[2]**
(b) • Brain death involves the death of both the cerebrum and brain stem
- The brain stem is important for the control of autonomic/involuntary responses/homeostasis
- The pupil reflex is controlled by the autonomic nervous system/is involuntary
- The iris contains antagonistic (circular and radial) muscles
- In bright light the circular muscles contract causing the pupils to constrict
- Pupils constrict as a result of the parasympathetic nervous system and dilate as a result of the sympathetic nervous system
- Pupils constrict to reduce overstimulation of the retinal cells/prevent damage to the retina
- In dim light radial muscles contract causing the pupils to dilate
- In dim light pupils dilate so there is enough light to see
- The absence of a pupil reflex shows brain stem death
- Without brain stem autonomic responses/homeostasis cannot occur
- The cerebrum controls higher order brain functions
- If the brain stem is functioning and the cerebrum is not the person is in a vegetative state
- A person in a vegetative state can carry out autonomic functions/homeostasis
- People may argue that the person is dead if they have lost higher order functions **[Max 4]**

4. (a) Innate **[1]**
(b) Bw: south east, Ub: south west **[1]**
(c) As the geographic distance increased the genetic distance between the species increased / positive correlation **[1]**
(d) • Interbreeding has still been able to occur
- The populations are not geographically isolated enough for speciation
- Gene flow can occur during migration and between geographically close species **[Max 1]**
(e) • Variation in the direction of migration due to a mutation
- Fr population have inherited a genetically based mutation/allele that has changed their migration pattern
- Increases in temperature have allowed the birds that fly north to survive
- The birds that fly north mate and pass on their alleles to their offspring
- The offspring inherit the allele for migrating north
- The allele frequency in the population has increased **[Max 3]**

5. • Damage to the brain/brain lesions/accidents/strokes cause loss of function of the damaged area and is compared with changes in behaviour
- However, damage may be across more than one area of the brain so it's difficult to interpret results
- Animal experiments involve directly stimulating the brain of live animals following removal of the skull allowing the targeting of specific areas of the brain

- However, the parts of animals' brains may function differently from humans / it may be considered unethical to cause suffering/ destroy the animal
- fMRI/functional magnetic resonance imaging looks at increased brain activity in response to particular stimuli
- Non-invasive technique
- Harmless dye is injected into the blood to show the blood flow
- It gives real-time results
- Increased brain activity is measured as increased blood flow / brain activity is shown in (false) coloured images
- fMRI scans can be used to analyse how healthy individuals' brains are
- fMRI scans can be used to analyse how healthy individuals respond to stimuli e.g. thought/perception/music/maths/ language/images
- fMRI links a specific stimulus to a specific area of the brain
- fMRI can be used to diagnose reduced oxygen flow to the brain/ strokes
- Disadvantage of fMRI is resolution of area of brain is poor as it looks at blood flow not neurons
- Not all brain activity shows up on MRI scans
- Resolution is poor (compared to PET scans) **[Max 6]**

Option B: Biotechnology and bioinformatics

1. (a) Biopharming **[1]**
   (b) • Antithrombin cannot be synthesized by bacteria as it is a glycoprotein so requires post-translational modification
   • The yield from rabbit milk would be too low **[Max 1]**
   (c) • Human gene for normal blood clotting antithrombin isolated
   • Gene is attached to milk regulatory elements so it is expressed in milk only
   • Microinjection of human gene into goat embryo
   • Gene regulated so expressed in goat milk
   • Antithrombin secreted in milk
   • Purified and used to treat antithrombin deficiency / prevent thrombosis **[Max 3]**
   (d) • Microarrays test for specific DNA sequences in healthy and diseased cells / microarrays detect the expression of multiple DNA sequences at the same time
   • mRNA is converted to cDNA by reverse transcriptase
   • DNA markers complementary to healthy DNA/cDNA are attached to green fluorescing markers, red markers are used for diseased cells, and yellow for both
   • DNA markers are attached to a surface
   • Patient's mRNA is converted to cDNA using enzyme reverse transcriptase
   • cDNA from the parents is exposed to the markers and then rinsed
   • If cDNA binds it will fluoresce
   • Red means gene is not expressed; green means gene is expressed **[Max 3]**

2. (a) • BLAST/Basic searches for the H1N1 gene/protein was compared with virus sequence that affects other species
   • Databases compared sequences by aligning nucleotides/ proteins and looking for similarities and differences
   • Sequences found to be most similar to US and UK swine flu genes **[Max 2]**
   (b) • RNA from cells is extracted from a person with influenza
   • The RNA contains both human and viral genes
   • Enzyme reverse transcriptase is used to make cDNA
   • Primer sequences for a specific strain of influenza are added
   • Primers cooled to 55 °C to allow the primers to join/anneal to the DNA
   • If the complementary sequence from the strain is present primer binds to cDNA
   • Taq polymerase/polymerase enzyme from bacteria *Thermus aquaticus* is added
   • Temperature increased to 73/74/75 °C to allow enzyme/Taq polymerase to replicate the strands of DNA
   • If the strain of influenza is present many copies of DNA are made
   • DNA containing the viral gene is tagged with fluorescent dye/ marker if present
   • If the sample fluoresces that particular strain is diagnosed **[Max 4]**

3. (a) Bioremediation **[1]**
   (b) Violet/purple **[1]**

   (c) • In batch fermentation the product is periodically collected whereas in continuous it is constantly collected
   • In batch fermentation the nutrients are added at the start whereas in continuous fermentation they are constantly added
   • In batch fermentation more frequent sterilization is required before the next batch whereas in continuous fermentation sterilization is less frequent **[Max 2]**
   (d) 79% **[1]**
   (e) • Optimum is pH 7/neutral
   • Sharply decreased under acidic <u>and</u> alkaline conditions / sharply decreases under extremes of pH **[1]**
   (f) • Both species reduce crude oil significantly under optimum conditions of pH <u>and</u> salinity / Both are most effective at 20g/L salt concentration <u>and</u> pH7/neutral
   • At pH 7/neutral pH *Bacillus cereus* (T-04) is more effective / degrades 79% whereas *Bacillus halotolerans* (1-1) degrades 65%
   • *Bacillus cereus* (T-04) has a higher degradation rate/ more effective at pH 5-7 / *Bacillus halotolerans* (1-1) more effective at pH 8
   • *Bacillus halotolerans* (1-1) more effective / at bioremediation of salty soils/soils 20g/L salt and above
   • Both / *Bacillus halotolerans* (1-1) and *Bacillus cereus* (T-04) are not as effective at extremes of pH of soil
   • May not be effective as is a simulation / both bacteria may be outcompeted by bacteria existing in soil **[4]**

4. **Advantages**
   • Vaccines offer cost-effectiveness
   • EASY to grow as just need soil
   • Easily administrable (plant seeds and grow in soil)
   • Easy to store / don't require refrigeration
   • Prevents infectious diseases
   • Cheap **[Max 3]**
   Disadvantages
   • Variation of dosage level within/between plants
   • Protein must not be digested
   • Length of immunity offered by vaccine may vary **[Max 3]**

Option C: Ecology and conservation

1. (a) (i) 33 – 18 = 15 **[1]**
   (ii) • Community of plants and animals that occupy a specific area /collection of ecosystems
   • With similar climate/rainfall and temperature **[2]**
   (iii) Tropical seasonal forest / Tropical savannah **[1]**
   (b) Tropical seasonal forest / Tropical savannah/Calcutta has a higher net primary productivity as it has a higher temperature **[1]**

2. (a) Producers/autotrophs **[1]**
   (b) • Symbiotic/mutualistic/endosymbiotic relationship
   • Both organisms benefit
   • Coral polyps provide shelter/nitrogen for the *Zooxanthellae*
   • *Zooxanthellae* photosynthesize and give sugars/oxygen/amino acids to coral polyp **[Max 2]**
   (c) • Increased temperatures / increased acidity of the oceans
   • Coral polyps eject the symbiotic *Zooxanthellae* causing them to bleach
   • If the coral dies fish decrease/reducing biodiversity
   • Calcium carbonate cytoskeletons dissolve **[Max 2]**
   (d) • Fish such as the parrot fish graze on algae that grow on coral
   • As a result, the algae population is kept under control/limited **[2]**

3. (a) • Indicator species are sensitive to a narrow range of environmental conditions / narrow tolerance of environmental conditions
   • Their population indicates a specific environmental condition **[Max 1]**
   (b) Total number of all individuals collected **[1]**
   (c) 2.6 **[1]**
   (d) B
   • As it has more pollution-tolerant organisms
   • As it has a lower biotic index **[2]**

4. (a) Autotroph **[1]**
   (b) • Waterlogging occurs, decreasing oxygen in the soil
   • Less nitrifying bacteria such as *Nitrosomas* / *Nitrobacter*
   • Increased denitrification occurs / *Pseudomonas*
   • Resulting in nitrogen-deficient soils

- Increase in carnivorous plants/sundew/Venus fly traps/pitcher plants
- Runoff from water leaches nitrogen into rivers/streams/ponds
- Causing eutrophication **[Max 3]**

(c) DNA/RNA/ATP **[1]**

(d) • Increased use of fertilizer/intensive farming
- Increased use of detergents **[Max 1]**

(e) Demand is increasing but phosphate from rocks is a limited resource and phosphates may run out **[1]**

5. • Populations increase when (natality + immigration) > (mortality + emigration)
- 'J'-shaped population curves occur when the population has an ideal environment/unlimited resources
- 'J'-shaped curves show continued exponential growth
- Exponential growth occurs when there are few predators/little disease/plentiful resources
- 'S'/sigmoid population curves occur when a population colonizes a new environment
- 'S'/sigmoid curve – initially resources are plentiful / few predators allowing exponential initial growth
- 'S'/sigmoid curve – exponential phase / 'J' exponential phase occurs when natality and immigration are high
- Transitional phase – competition for resources/space
- Transitional phase – population growth increases but at a slower rate
- Plateau/stationary phase – the population stops growing when it reaches the carrying capacity
- Carrying capacity is the maximum population size that can be supported by an environment
- A population reaches carrying capacity when natality and mortality are equal / (natality + immigration) = (mortality + emigration)
- The plateau phase oscillates due to predator–prey interactions **[Max 6]**

Option D: Human physiology

1. (a) • NVP reduces the blood glucose levels compared with the control
  - 5 g NVP halves the blood glucose levels compared with the control
  - 2.5 g NVP reduces the blood glucose level compared with the control but reduction is not as much as 5g NVP
  - 5 g is significantly more effective than 2.5 g as the error bars don't overlap **[Max 2]**

(b) • It increases the bulk of the food so less glucose is absorbed
- Glucose is absorbed more slowly / levels do not spike as high **[Max 1]**

(c) • Type II diabetes
- Bowel cancer / haemorrhoids / appendicitis
- Strokes
- CHD / coronary heart disease / hypertension **[Max 2]**

(d) • Fibre absorbs water adding bulk to feces
- Feces travel through the large intestine faster
- Prevents constipation **[Max 1]**

(e) • Decreased appetite in response to the hormones, leptin, PYY3-36 and insulin
- When the small intestine contains food, it secretes the hormone PYY3-36
- Adipose tissue secretes the hormone leptin when more fat is stored
- The pancreas secretes the hormone insulin in response to high blood sugar **[Max 2]**

2. (a) • Type II pneumocyte
- As it is a large cell with a cuboidal shape / small surface area to volume ratio / not thin and long / contains lots of rough endoplasmic reticulum / Golgi apparatus/contains secretory granules / many microvilli **[2]**

(b) • Surfactant
- To prevent the alveoli from collapsing / reduce surface tension **[2]**

3. (a) • P: Atrial depolarization/contraction/systole
- QRS: Ventricle depolarization/contraction/systole
- T: Ventricle repolarization/diastole **[3]**

(b) 4 squares × (0.2/5) = 0.16 s **[1]**

(c) • Allows the blood to flow from the atria/atrial systole before the AV/atrioventricular valves shut / ventricular systole **[1]**

(d) • Lup – mitral valve/atrioventricular / bicuspid/tricuspid valves close

- Dup – semilunar valve closes **[2]**
  [Award 'valves close' if no other mark]

4. (a) 30 mmHg **[1]**

(b) • In the lungs there is little difference/only 5% difference between the saturation of oxy-hemoglobin in red blood cells at all CO_2 concentrations
- In the muscle tissues the greater the concentration of CO_2 the lower the percentage saturation of oxy-hemoglobin **[2]**

(c) • It shifts to the left
- Due to an increase in CO_2 concentration from respiration **[2]**

5. • Hormones involved are prolactin and oxytocin
- A suckling baby sends impulses from the sense receptor in the nipples / along a sensory neuron to the hypothalamus
- The hypothalamus releases neurochemicals/PRH/ prolactin releasing hormone into portal vessels
- Neurochemicals act on the anterior lobe/adenohypophysis of the pituitary to release prolactin
- Prolactin stimulates the development of the mammary glands
- Prolactin stimulates the synthesis of milk from mammary glands/ alveoli
- The hypothalamus contains neurosecretory cells that synthesize oxytocin
- Oxytocin is released by the posterior lobe/neurohypophysis of the pituitary
- Oxytocin stimulates the release of milk
- Positive feedback loop between oxytocin and hypothalamus and pituitary
- Progesterone inhibits prolactin results during pregnancy **[Max 6]**

SET B

Set B: Paper 1

Question no.	Answer	Question no.	Answer	Question no.	Answer
1.	C	15.	B	29.	C
2.	A	16.	B	30.	B
3.	C	17.	B	31.	B
4.	A	18.	B	32.	C
5.	A	19.	B	33.	C
6.	C	20.	C	34.	D
7.	C	21.	B	35.	A
8.	C	22.	D	36.	B
9.	A	23.	D	37.	A
10.	A	24.	B	38.	A
11.	D	25.	B	39.	B
12.	B	26.	A	40.	D
13.	C	27.	B		
14.	D	28.	C		

Set B: Paper 2

Section A

1. (a) (i) 230 nA
   (ii) 160 nA **[1 for both]**

(b) (230 – 160) ÷ 230 × 100 = 30% decrease (ECF) **[1]**

(c) 86/87/88 mV **[1]**

(d) **Similarity**:
- Both depolarize 6/7 seconds after stimulation
  Differences:
- The HD mice maximum membrane potential/depolarization is slightly lower/similar to the control / no significant difference / numerical values accurately read
- The time taken for repolarization in Huntington's disease/HD is longer than in control/WT **[2]**

(e) • Repolarization
- Depolarization height/times are similar
- Repolarization takes longer in HD / acceptable numerical comparison
  [Max 2; 1 mark for incorrect answer with correct reasoning]

(f) (i) Open
   (ii) Closed **[1 for both]**

(g) The more negative the voltage (inside the fibre) the greater the current flow into the axon (through potassium channels) **[1]**

(h) Huntington's disease decreases current flow
- Smaller current for every HD throughout
- Differences are statistically significant
- HD current is approximately 1/3 that of control for chloride channels

- −140mV is −360I_{Cl} μA/cm² for HD and −1133 I_{Cl} μA/cm² for WT
- HD current is approximately 1/3 that of control for potassium channels (at −60 mV) / −87 I_{Cl} μA/cm² for HD and −216 I_{Cl} μA/cm² for WT **[Max 2]**

(i) 7a⁺ / spliced mRNA / no exon 7a **[1]**

(j)
- High 7a⁺ mRNA results in lower relative expression of Clcn1 mRNA
- HD/Huntington's disease mice have three times the amount of 7a⁺ mRNA / numerical values are 18% HD vs 6% WT
- HD expression of the Clcn1 mRNA is about 20%/0.2/five times less than WT expression **[3]**

(k)
- Huntington's disease muscle fibres require a lower stimulus current to trigger depolarizing/action potential / Huntington's disease muscle fibres are hyperexcitable/more easily triggered to contract (first graph)
- Huntington's disease muscle fibres take longer to repolarize following an action potential/action potentials / contraction lasts longer in Huntington's disease (second graph)
- Huntington's disease results in reduced peak current (I_{Cl} μA/cm²) flowing through chloride and potassium voltage-gated channels during repolarization
- The expression of muscle chloride channel/ClC-1 in Huntington's disease is reduced
- Huntington's disease contains an incorrectly spliced mRNA/7a⁺ leading to a fall in functioning chloride channels (in Huntington's disease skeletal muscle)
- Huntington's disease results in a faulty protein/chloride channel **[Max 3]**

2. (a) Interphase, as nuclear membrane is visible / DNA has not condensed **[1]**

(b)
- Undifferentiated but have the ability to differentiate / multipotent/totipotent/pluripotent
- Unspecialized but have the ability to specialize
- Have the ability for repeated cell division/mitosis **[Max 2]**

(c)
- The cell cycle consists of interphase/G1 S G2 and mitosis
- 4 Cyclins/ Cyclin A D E K regulate the cell cycle
- The concentrations of cyclins increase and decrease throughout the cycle
- The change in concentrations of cyclins stimulate the next phase/timing of the cell cycle
- Cyclins bind to cyclin-dependent kinases to become activated
- Kinases phosphorylate (target) proteins to activate them
- Phosphorylated proteins have a particular function in the cell cycle / example of function e.g. duplicate
- After the event has happened the cyclins become inactive again
- Example – Cyclin D stimulates the G1 phase/interphase / Cyclin E stimulates G1 stage to move to S phase / Cyclin A activated DNA replication/causes S phase to move to G2, Cyclin B causes G2 phase to prepare for mitosis/assemble mitotic spindle **[Max 3]**

(d) One cause from:
- Mutation in gene in retina / causes protein involved in active transport to degenerate
- Results in vision loss

One treatment:
- Source of stem cells – embryonic/embryos
- Treated by replacing dead cells in retina with stem cells **[Max 2]**

3. (a) Series of enzyme-controlled reactions organized into chains or cycles / anabolism/catabolism **[1]**

(b) Non-competitive inhibition **[1]**

(c)
- Slows reaction rate as (non-competitive) inhibitor binds to the allosteric site
- Causes the shape of the active site to be changed
- So the active site cannot bind to the substrate
- Binding of the inhibitor is irreversible
- The maximum rate of reaction is lower with the inhibitor than without even at high substrate concentrations **[Max 2]**

4. (a) Gene/autosomal linkage/linked genes **[1]**

(b) **[1 mark for 2 correct gametes] [2]**

c S	C s	c s	c S

(c)

C s	c S
c s	c s

[1]

(d)
- Yes, the observed differs significantly from the expected ratio / The expected ratio is 1:1:1:1/25% for each
- Three degrees of freedom/3df

- Critical value is 7.81 (for P = 0.05) **[Accept 11.34 for P=0.01 *or 16.27 for P= 0.001*]** / circled value
- Chi-squared value/X^2 value is greater than critical value
- So are 95% confident (99% confident / 99.9% confident based on critical value * for P=0.01 or P=0.001)
[Allow ECF following an incorrect degree of freedom] [Max 2]

(e)
- Crossing over between non-sister chromatids/chromatids on homologous chromosomes
- In meiosis I/prophase I
- Causes exchange of alleles / chromosomes exchange sections / crossing over takes place
- Chiasmata form between (non-sister) chromatids
- Sections have different alleles / new combinations of (linked) alleles / new linkage groups **[Max 2]**

(f)

Angiospermophytes	Bryophytes
Flowers	No flowers
Seeds	Spores
Roots	No <u>true</u> roots / rhizoids
Leaves	No true leaves / thallus
Waxy cuticle	No waxy cuticle
Vascular tissue / xylem and phloem	No vascular tissue / no xylem or phloem

[Max 2]

Section B

5. (a)

Genes	Short tandem repeats
Not repetitive sequence of bases (e.g. ATCGATATATATACG)	Highly repetitive sequence of bases (e.g. TATACTATACTATACTACAC)
Code for a polypeptide/ protein	Do not code for a protein
Unique in genome	Occurs many times in genome
Exons	Introns
Translated	Not translated
Similar among individuals with same allele	A lot of variation between individuals
Not used to create a DNA profile	Used to create a DNA profile
Eukaryotes and prokaryotes	Eukaryotes

[Max 3]

(b)
- DNA extracted from the nucleus of a cell e.g. hair, cheek, blood, semen, tooth, bone
- <u>PCR/polymerase chain reaction</u> to amplify small amounts of DNA / produce large amounts of DNA
- Short tandem repeats / repetitive non-coding sections of DNA are used
- <u>Restriction</u> enzyme/endonuclease cuts DNA at a specific sequence resulting in different size fragments
- <u>Gel electrophoresis</u> involves using electricity to separate fragments
- DNA has a negative charge so migrates to the positive terminal
- Small fragments travel further than large fragments
- Resulting in bands unique to an individual
- **Application** – bands can be used:
  o To detect paternity by comparing/matching to a father
  o In forensics to determine if a person was at a crime scene / compare victim's DNA to DNA on belongings to identify the victim **[Max 4]**

(c)
- Evolution is the change in inherited/heritable characteristics of a species over time
- DNA is a very stable molecule / adenine pairs with thymine and cytosine with guanine
- The DNA code is universal / the same codon codes for the same amino acid in every species
- Genes/proteins can be compared between organisms (e.g. cytochrome c/mitochondrial DNA/hemoglobin/ATP)

- Mutations occur at a constant rate / mutations accumulate over time
- Less similar organisms have more differences in their base sequence
- The greater the difference in base sequence the longer ago the species diverged from a common ancestor
- The number of mutations can be used to determine the <u>time</u> the species diverged from a common ancestor / molecular clock
- Less subjective/more objective/more accurate than using observable characteristics to classify organisms
- Changes in DNA result in changes in amino acid sequences, which can also be used as molecular clocks
- DNA molecular clocks are more accurate than amino acid molecular clocks as the genetic code is degenerate / mutations in DNA do not always result in different amino acids as more than DNA sequence codes for an amino acid
- Mutations are chance events so more than one molecular clock should be used **[Max 8]**

6. (a)
- Antibodies/immunoglobulins kill pathogens by lysing the cell wall / perforin
- Antibodies bind/tag pathogens which then attract phagocytes/macrophages / opsonization
- Antibodies cause agglutination/sticking together of pathogens, preventing pathogens entering cells
- Antibodies bind to viral docking proteins, preventing the virus from entering cells / neutralize viruses
- Antibodies can neutralize toxins / act as antitoxins
- Antibody-coated antigens stimulate the complement cascade that causes cell walls to lyse/burst **[Max 3]**

(b)
- Circadian rhythm controls sleep/awake cycles
- Hormone melatonin
- Controls daily sleep/wake cycle
- Produced in pineal gland
- Acts on hypothalamus...
- ...in the suprachiasmatic nuclei
- Melatonin secretion increases at dusk/decreases at dawn
- Melatonin secretion causes drowsiness/enables sleep
- Melatonin is broken down by the liver
- Negative correlation between age and melatonin secretion / older people secrete less melatonin than younger people
- Jet lag is caused by disturbance in time zones/flying east on sleep–wake cycle
- Reduced by artificially taking melatonin when person wants to sleep **[Max 6]**

(c)
- <u>Vaccine is a weak/attenuated form of the virus</u>
- <u>Antigen</u> on membrane of pathogen / <u>antigen</u> is displayed on membrane of a phagocyte/antigen-presenting cell
- T cells/T$_H$ cells/T lymphocytes bind to the antigen
- T cells/T$_H$ cells/T lymphocytes release cytokines / activate specific B cells/B lymphocytes
- B cells/B lymphocytes divide by mitosis / produce clones
- Forming plasma cells/B cells that release antibodies
- Antibodies bind to virus antigens causing destruction of virus/agglutination/cell lysis/precipitation/perforation/neutralization
- Some B cells make memory cells
- Memory cells make antibodies faster/more antibodies upon exposure to pathogens **[Max 6]**

7. (a) Diagram could include:
- Cortex (shown as thinner layer than medulla at convex side of kidney)
- Medulla (shown as pyramids between cortex and pelvis)
- Pelvis (shown connected to medulla and ureter on concave side of kidney)
- Ureter (shown connected to pelvis)
- Renal artery and vein (connected to top concave side of kidney)

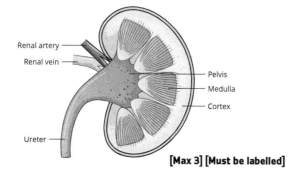

[Max 3] [Must be labelled]

(b)
- Renal artery has more glucose as glucose is used by the kidney in respiration / to give energy for active transport
- Renal artery has more oxygen as oxygen is used by the kidney in aerobic respiration / to give energy for active transport
- Renal vein has more carbon dioxide as it is produced from respiration / waste product of metabolism
- Renal vein has less urea as it has been filtered but not reabsorbed / excreted
- Renal vein has less toxins as they have been excreted
- Protein levels are the same as protein is not filtered
- Water and sodium chloride levels may differ due to osmoregulation **[Max 4]**

(c)
- Nitrogenous waste is dissolved in the haemolymph as ammonia
- Ammonia is toxic
- Ammonia (NH$_4$) is <u>actively</u> transported
- Ammonia (NH$_4$) is transported from the <u>haemolymph</u>...
- ...into the <u>Malpighian</u> tubule system
- Ammonia (NH$_4$) is converted into less toxic <u>uric acid</u> by (Malpighian tubule) cells
- Sodium ions/Na$^+$ <u>and</u> chloride ions/Cl$^-$ ions are actively transported out of the haemolymph/into the Malpighian tubules
- Increased solute concentration in Malpighian tubules
- Water enters the Malpighian tubules by <u>osmosis</u> and flushes uric acid into the hind gut
- Sodium ions/Na$^+$ <u>and</u> chloride ions/Cl$^-$ ions are <u>actively</u> transported out of the feces/rectum into the haemolymph
- The haemolymph becomes hypertonic / increases in solute concentration
- Water leaves the feces/rectum by osmosis **[Max 8]**

Set B: Paper 3

Section A

1. (a)
- Surface area/size of cube
- Time soaked for **[Max 1]**

(b) Potato – 0.32 mol dm^{-3} and sweet potato – 0.47 mol dm^{-3} **[1]**

(c) Any value between 0.32 and 0.47 **[1]**

(d)
- Potato – 0.64 osmoles per litre/Osm/L
- Sweet potato – 0.94 osmoles per litre Osm/L **[2]**

(e)
- No repeats for each interval
- Temperature not controlled
- Potato/sweet potato not blotted to remove excess water **[Max 1]**

(f) Sweet potatoes contain significantly/lots more sugar per 100 g so they have a higher isotonic point **[1]**

2. (a) X = prophase **[1]**

(b)
- Scale bar image size = 18 mm = 18,000 µm
- Diameter = 25,000 µm
- Actual size = 25,000/18,000 × 10
- 13.9 µm (accept between 13 and 15 µm) **[1]**

(c) X = 62% and Y=27% **[1]**

(d) 1 – the meristem is where stem cells / rapidly dividing tissue are found / has a higher mitotic index **[1]**

3. (a)
- Only proteins contain sulphur in disulphide bridges
- Only DNA contains phosphorus in the sugar–phosphate backbone **[1]**

(b) Carbon is present in both DNA and proteins so would not differentiate between them **[1]**

(c)
- DNA is the genetic material
- The radioactive DNA was passed on to daughter cells
- Sulphur is not the genetic material as the radioactive sulphur was not passed onto daughter cells **[2]**

Section B

Option A: Neurobiology and behaviour

1. (a) Cerebral cortex **[1]**

(b)
- (Cerebral cortex) is highly developed in humans for perception/awareness/memory/thought/language/consciousness
- (Cerebral cortex) is larger/thicker in humans
- (Cerebral cortex) occupies a higher proportion of the brain in humans
- (Cerebral cortex) has a larger surface area in humans
- (Cerebral cortex) extensively folded in humans to accommodate the larger surface area/ fit brain in skull **[Max 1]**

(c)
- Brain metabolism is linked to glucose uptake / the larger the brain the more neurons it contains
- More glucose is required due to increased number of neurons
- Neurons require energy from respiration

- Energy is required to activate sodium–potassium / Na+/K+ pumps / in repolarization of neurons
- Energy is required to synthesize neurotransmitters
- Primate brains respiring more than rodent brains so require more glucose
- Human brains have the highest amount of respiring cells/ neurons **[Max 2]**

(d)
- Memory is the process of encoding storing, and retrieval of acquired information
- Encoding is how the brain converts information into a way it can be stored
- Retrieval of information is the ability to recall the memory
- Short-term memories are quick to form and last for a short time
- Some short-term memories lead to long-term memory which are stored for a long time
- The hippocampus makes many new connections/dendrites during the formation/encoding of a memory
- The neurons are then pruned for more efficient recall **[Max 3]**

2. (a)
- Sound stimuli increase activation in the visual cortex
- Sound stimuli have less activation in the visual cortex than visual stimuli **[Max 1]**

(b)
- Plasticity increases dendrites making more synaptic connections / neural pathways
- The visual cortex adapts to process auditory rather than visual stimuli
- The person can 'see' sound **[2]**

(c)
- If the hair cells in the cochlea are damaged hearing aids will not work
- Cochlea implants allow deaf people to hear sounds but not clearly
- Cochlea implants have an external microphone that receives the sound
- A processor filters out frequencies of sound not in the normal speech range
- A transmitter sends the sound to the internal implant receiver
- The internal part of the implant is embedded in bone behind the ear
- The receiver detects the sound and a stimulator converts sound energy to electrical impulses
- The electrical impulses pass along an electrical array to stimulate the auditory nerve **[Max 4]**

3. (a) Negative phototaxis **[1]**

(b) An inherited instinctive response that is not influenced by learning/environment **[1]**

(c)
- Yes, because there is a significant difference between the observed frequencies and expected frequencies
- The null hypothesis is rejected, and the alternative hypothesis is accepted
- As the value of chi-squared is higher than the critical value of 7.82
- For 3 degrees of freedom
- The chi-squared test is valid as all categories have more than 10 **[3]**

4. (a)
- Relatedness is not the strongest motivator / both allogrooming and food donated are stronger motivators than relatedness, which has the steepest line / relatedness is the weakest motivator
- The strongest motivator for donating food is the amount of allogrooming received **[2]**

(b)
- Reciprocal altruism/learned behaviour
- Bats do not starve so reproduce and pass on their genes
- Increasing the population of the bat colony **[2]**

(c)
- Other males would compete for mates
- Reducing the chance of the male reproducing and passing on his genes / reducing the male's fitness in the population **[1]**

5.
- Operant conditioning is a method of learning that occurs through rewards or punishments from behaviour
- An association between the behaviour and the consequence of the behaviour is learned
- It occurs through trial and error
- Reinforcement increases the chance of the behaviour, punishment decreases the chance of the behaviour
- Example of positive reinforcement – hungry rats in a box got food if they accidently pushed a lever
- Example of negative reinforcement – loud noise stopped if rats pushed a lever
- Example of positive punishment – electric floor shocks rat after lever is pressed

- Rats learned to associate pushing the lever with the food/ silence/shock
- The operant response was pressing the lever
- The food/noise/shock is a reinforcement of the behaviour
- Loud noise stops when a lever is pressed is negative reinforcement **[Max 6]**

Option B: Biotechnology and bioinformatics

1. (a) **Advantage**
- Hairless so fluorescing easier to see
- Hairless so more similar to humans
- No immune system to ensure tumour growth
- Could lead to improvements in medicine for people
- It is unethical to test on people before it has been shown to be effective on animals
- The mouse is a frequently used model organism so gene function in mice can be a good predictor in humans

Disadvantage
- Mice may not respond in the same way as humans
- Ethical issues **[Max 4]**

(b) Using florescent dye as image is sharper / higher resolution image **[1]**

(c) Transferrin **[1]**

(d) Vehicle mouse as it has greater fluorescence (also accept control mouse) **[1]**

(e)
- The higher the fluorescing the greater the tumour size
- Sorafenib/40 mg/kg mouse and the control/vehicle mouse have no significant difference in the tumour size for the first 2 days
- At 3 days the mouse containing sorafenib tumour had shrunk by approximately half
- Not effective in stopping tumour growth as after 10 days the tumour had grown significantly/doubled in size
- Effective at reducing the speed of tumour growth as the tumour doubled in size but the control mouse tumour tripled in size
- A higher dose may cause a greater reduction in tumour growth **[Max 4]**

2. (a)
- Some genetic diseases result in the presence of abnormal metabolites
- The abnormal metabolites can be detected in the blood
- Early diagnosis is important as some can be treated before symptoms appear
- If level of phenylpyruvate is high the baby has PKU
- Untreated PKU results in impaired intellectual function/ seizures
- If detected early changes in diet prevent severe symptoms for PKU
- Treated by a low phenylalanine diet **[Max 3]**

(b)
- Abnormal metabolites / trypsin / target molecule can be detected using enzyme-linked immunosorbent assay
- Can be used in diagnosis by identifying abnormal metabolites / trypsin / target molecule from saliva/plasma/blood
- Capture antibody/molecule is immobilized in a well plate
- Capture antibody/molecule is complementary to trypsin / binds to trypsin
- The well is rinsed so any unbound antigens/trypsin is washed away
- A second monoclonal antibody bonded to an enzyme is added that binds to the abnormal metabolites / trypsin / target molecule
- The well is rinsed to remove any unbound abnormal metabolites / trypsin / target molecule
- A solution containing a substrate is added
- If the enzyme is present the substrate will change colour indicating the presence of the antibody
- If the antibody is not present, there will be no colour change as there would be no antibody for the secondary antibody to bind to
- The greater the colour change the greater the amount of antibodies **[Max 3]**

3. (a) The use of microorganisms to remove an environmental pollutant from soil or water **[1]**

(b)
- Oil spills result in macroscale pollution / float on oceans
- Hydrocarbons persist in the environment for a long time
- Hydrocarbons are biodegradable using *pseudomonas bacteria*
- Pseudomonas use hydrocarbons as their source of energy / are chemoheterotrophs
- Can break down toxic compounds in oil/hydrocarbons to non-toxic products / carbon dioxide and water

- Growth of bacteria is enhanced by addition of nutrients
- Physical methods such as pumping oil out of water are expensive and slow
- Chemical methods can result in bioaccumulation / biomagnification in tissues as they are toxic and are not biodegradable **[Max 5]**

4. (a) 8,497 base pairs **[1]**
   (b) DNA as it contains T thymine / DNA as, if it was RNA, it would contain U **[1]**
   (c) • The DNA in the actin gene is the same as there is 100% match / 8,497/8,497 bases match suggesting they could be the same species
      • However, this is only looking at one gene so is not enough to determine if it is the same species
      • More genes need to be looked at to determine if they are the same species **[Max 2]**
   (d) • Nucleotide sequences are better as mutations in bases are always identified, but those in proteins are not due to degenerate nature of proteins. **[1]**
   (e) • No, as plants do not contain actin
      • Actin is a fibrous protein found in animal muscle
      • To look at differences in plant phyla proteins that are present in plants are needed **[Max 2]**

Option C: Ecology and conservation

1. (a) • Transect placed from mid- to high shore / up the shore
      • Optic level was used to measure height above chart datum
      • Every 0.5 m above chart datum the number of barnacles was counted using a quadrat **[Max 2]**
   (b) • Increased competition with space / interspecific competition
      • Increased predation **[1]**
   (c) • Semibalanus balanoides as it is not present on the upper shore
      • Semibalanus balanoides as it is emersed for longer **[1]**
   (d) The principle of competitive exclusion states that when two species compete for exactly the same resources/occupy the same niche then one will outcompete the other
   At 2.5–4.0 m /lower mid-shore to middle mid-shore Semibalanus balanoides outcompetes both species of Chthamalus as it is in ideal conditions and outcompetes the others for space
   At 4.0–5 m / upper mid-shore to lower upper shore Chthamalus stellatus outcompetes Semibalanus balanoides as it is better at resisting desiccation but it outcompetes Chthamalus montahui for space
   Above 5 m Chthamalus montahui is the most common species as it can resist desiccation the best
   The fundamental niche of Semibalanus balanoides is increased competition with algae for space / interspecific competition **[Max 3]**
   (e) • The Chthamalus montahui has a fundamental niche between 2.5 and 6 m
      • The realized niche is 4.5-6 m
      • Due to being outcompeted by Chthamalus stellatus at 4.5 m
      • Due to being outcompeted by Semibalanus balanoides below 4 m **[Max 2]**

2. (a) Any one of the following for [1] mark:
      • A top-down factor refers to predation/herbivory/a higher trophic level influences the population of a lower trophic level whereas bottom-up refers to availability of nutrients / lower trophic level influencing the population of a higher trophic level.
      • In top-down control an increase in higher trophic level/ predator population causes a decrease in the lower trophic level/prey population, whereas in bottom-up an increase in lower trophic level/prey population causes an increase in the higher trophic level/predator population size.
      • In top-down control a reduction in the population size of one trophic level increases the population size of the next which reduces the next, whereas in bottom-up a reduction of the lower trophic level reduces the population size of all higher trophic levels.
      • A top-down control example is keystone species/apex predators controlling the population sizes of lower trophic levels. Example of bottom-up is deforestation/nutrient depletion reducing food for higher trophic level. **[Max 1]**
   (b) Primary producer (autotrophs) **[1]**
   (c) • Bottom-up forcing/interaction
      • E.g if kelp increased the sea urchin populations would increase as there's more food so sea otter populations increase as more prey

- E.g if kelp increased the sea urchin population would increase as they have more food so the gull population would increase as there is more food, which would increase bald eagle populations as there is more food
- E.g if kelp populations increased the coastal fishes population would increase a there is more food, which would increase bald eagle population as they have more prey
- E.g if kelp increased the coastal fishes population would increase so the gull population would increase, as they have more prey, which would increase the bald eagle population
- E.g if kelp increased the musses/barnacles populations would increase so the starfish would increase as they have more prey, which would increase the sea otter population as they have more prey **[Max 2]**
   (d) • A keystone species has a disproportionate effect on an ecosystem compared with the population size
      • The absence of the keystone species causes the ecosystem to change dramatically **[Max 1]**
   (e) Kelp would decrease as more sea urchins would graze on the kelp **[1]**

3. (a) • Desert has a net primary productivity of 100 g cm^{-2} yr^{-1} whereas the tropical rainforest is 1,000 g cm^{-2} yr^{-1}
      • The tropical rainforest net primary productivity is 10 times higher than the desert
      • The tropical rainforest has greater variation in the net primary productivity than that of the desert **[Max 2]**
   (b) Rainforests have a higher net primary productivity as they have greater precipitation **[1]**
   (c) • Rainforests, as there is greater primary productivity / biomass/energy in first trophic level
      • Greater energy/biomass in first trophic level can support longer food chains
      • Rainforests have 10 times the energy in the first trophic level so they can support an extra trophic level (as 10% of the energy is transferred at each trophic level) **[Max 1]**

4. (a) 5 m: 19 and 10 m: 24 **[both needed for 1]**
   (b) The relative abundance of each species in an area **[1]**
   (c) 14.7. OR 14.72 **[1]**
   (d) 10 m depth **[1]**
   (e) Competition with other corals / herbivory / disease / parasitism **[1]**
   (f) Tropical rainforest

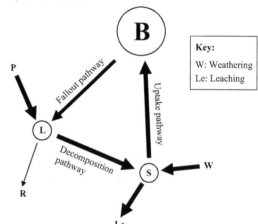

Desert

[Marking points below given on correctly drawn Gersmehl diagram]
- Main store of nutrients in desert is soil whereas in tropical rainforest it is biomass / S store/circle bigger in desert than rainforest

- Nutrients stored in biomass is <u>higher</u> in rainforests (due to dense vegetation and rich biodiversity) / B store/circle bigger in rainforest than desert
- Soil nutrient store in rainforest is lower in nutrients (due to rapid uptake by biomass/increased leaching of water) / smaller soil circle in rainforest diagram
- Rapid transfer between stores and environment in rainforest / thicker arrows in rainforest
- More decomposers in rainforest as it is moist and warm
- Litter to soil <u>and</u> soil to biomass is faster in rainforest
- Leaching occurs in rainforest but not in desert (as water dissolves soil minerals)
- Precipitation is greater in tropical rainforest / the P input arrow is thicker
- Runoff only occurs from the soil / occurs in rainforest (as not all water is absorbed by soil)
- Deserts can be cooler than rainforests
- Rainforests have higher annual rainfall / 250+ cm whereas deserts are below 100 cm **[Max 6. Max 4 without Gersmehl diagram]**

Option D: Human physiology

1. (a) P wave indicated as below **[1]**

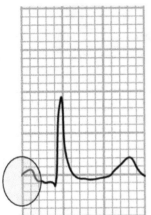

   (b) • From one R to the next, there are 6 squares
   - Each square is 0.2s meaning 1.2s for one beat.
   - 60s/1.2s = 50 beats per min **[1]**
   (c) SAN/sinoatrial node **[1]**
   (d) Anorexia nervosa = 2,818.9 mL / 2.8 L
   Control = 4,496.0 mL / 4.5 L **[2][If no units – max 1 mark]**
   (e) 97 − 71 g = 26
   (26/97) × 100 = 26.8%/27%
   [Accept any answer between 25% and 27%][1]
   (f) • The ventricle wall is muscle/protein
   - Anorexics have low fat and carbohydrates
   - Protein is being digested/respired for energy **[Max 1]**
   (g) • (Anorexia) lowers both systolic <u>and</u> diastolic blood pressure / Lowers mean blood pressure
   - Lower systolic blood pressure as the wall of the left ventricle is thinner
   - Lowers diastolic blood pressure as the cardiac output is lower / blood flow is reduced **[Max 2]**

2. (a) 61.9%
   Mother's blood = 2.1 mL O_2 /100 mL blood
   Fetal Blood = 3.4 mL O_2 /100 mL blood
   Percentage difference = (3.4 − 2.1)/2.1 × 100 = 61.9% **[1]**
   (b) • Fetal hemoglobin has a higher affinity for oxygen / binds to oxygen more easily than maternal hemoglobin
   - So oxygen will always diffuse from the maternal blood to the fetal blood in the placenta **[2]**
   (c) Four polypeptide chains containing 4 heme/iron groups, each bound to an oxygen molecule/four oxygen molecules in total **[1]**

3. (a) X = type II pneumocytes
   Y = type I pneumocytes **[1 mark for both correct]**
   (b) Emphysema as there are alveoli walls that have collapsed / large alveoli / large air spaces **[1]**
   (c) **Causes**
   - Irritants/chemicals in cigarette smoke
   - Increase in phagocytes increases inflammatory response
   - Inflammation releases the enzyme elastase
   - Elastase digests elastin in the alveoli walls reducing the surface area to volume ratio / reducing gas exchange
   - Can be due to a genetic mutation / inherited

Treatments
- Oxygen therapy
- Bronchodilators to dilate airways in lungs
- Corticosteroids/steroids
- There is no cure
  [Max 4][Max 3 if only causes or treatments given]

4. (a) X – Kupffer cell **[1]**
   (b) • Excess bilirubin in the blood
   - Due to blocked bile duct/inflamed bile duct / pancreatic cancer
   - Failure of the hepatocytes to break down bile / liver inflammation
   - Fast destruction of red blood cells / haemolytic anemia
   - Causes skin/eyes to look yellow / hyperbilirubinemia
   - Dark urine/pale feces **[Max 4]**
   (c) • Kupffer cell is engulfed and digests red blood by phagocytosis
   - Into heme and globin/protein
   - The globin/protein is broken down/hydrolysed/digested by peptidases/enzymes into amino acids
   - Heme is broken down by Kupffer cell into iron and bilirubin
   - Iron is transported in the blood by transferrin protein to the bone marrow
   - Bone marrow uses iron in newly formed new erythrocytes/red blood cells
   - Excess iron is stored in the liver as ferritin
   - Bilirubin is converted to bile
   - Bile is stored in the gall bladder **[Max 6]**

Set C

Set C: Paper 1

Question no.	Answer	Question no.	Answer	Question no.	Answer
1.	B	15.	A	29.	A
2.	B	16.	A	30.	B
3.	D	17.	B	31.	B
4.	B	18.	C	32.	A
5.	B	19.	C	33.	D
6.	D	20.	C	34.	A
7.	C	21.	B	35.	B
8.	B	22.	B	36.	C
9.	C	23.	D	37.	D
10.	B	24.	B	38.	D
11.	A	25.	B	39.	D
12.	D	26.	C	40.	A
13.	B	27.	C		
14.	D	28.	B		

Set C: Paper 2

Section A

1. (a) (4,613 ÷ 125,048) × 100 = 3.689%/3.69%/3.7% **[1]**
   (b) Epidemics/disease/virus that occur simultaneously in many countries / globally / across the world at the same time **[1]**
   (c) • Highest death rate is MERS-CoV (quantified) with 3 times the death rate SARS and 10 times the death rate of CoV-19
   - Greatest number of countries affected is CoV-19 with approximately 4/4.5 times the number of affected countries
   - Most total number of deaths caused by CoV-19 / Death rate is 5 times higher than MERS-CoV and 6 times higher than SARS-CoV
   - Data for CoV-19 are not complete as outbreak is current so total figures are not reliable **[Max 3]**
   (d) 3 March 2020: 2,000   11 March 2020: 10,000 **[1]**
   (e) (10,000−2,000) ÷ 2,000 × 100 = 400% **[(1) allow ECF from (a)] [1]**
   (f) • Fewer new cases as people who had the disease have died or recovered
   - More people were immune to the virus
   - Improved education/better awareness/ improved hygiene/ precautions/hand washing
   - Better treatments in hospitals
   - More hospitals/treatment centres set up/more doctors/ nurses/healthcare assistants
   - Use of antiviral drugs/new medicines
     [Do not accept antibiotics]
   - Better diagnosis leading to identification of infected and isolation
   - Less contact with animals that spread the disease to humans
     [1]

(g) • Not all people who have CoV-19 have been diagnosed so the death rate could be lower/some people have very mild symptoms/only test those presenting symptoms
  • Not enough capacity to test everybody for an accurate picture
  • Not all infected people who will die have died by 12 March as it's an ongoing epidemic so the final death rate could be higher **[Max 1]**

(h) • Start of epidemic came first in China/later in Italy
  • Peak of epidemic in China is earlier than in Italy/ China epidemic peaked on 4 February whereas Italy is highest on 12 March / not yet peaked by 12 March
  • Decrease in new cases in China but no decrease in new cases in Italy / China shows epidemic slowing down whereas in Italy it is still increasing
  • Higher maximum number of cases in China / converse for Italy
  • The rate of increase of new cases is similar in both countries
  • The death rate in Italy is higher/nearly double than China on 12 March
  • Italy's death rate of 6.6% is higher than the global death rate of 3.7% / China's death rate of 3.6% is similar to the global death rate of 3.7%
  • Large fluctuations in both countries **[Max 3]**

(i) • International cooperation allows sharing of knowledge/ collaborative research
  • Increases the chance of developing a vaccine
  • Increases the chance of developing antiviral drugs
  • Allows monoclonal antibodies to be made for diagnosis
  • Enables comparison of viral DNA to pangolin DNA to determine which animal originally transmitted the disease to humans **[Max 1]**

(j) • Wearing masks
  • Social distancing
  • Movement of air
  • Vaccination **[Max 1]**

2. (a) Anaphase II / two / 2 **[1]**
  (b) • Non-disjunction
    • Failure of separation of homologous chromosomes/ pairs of chromosomes (in anaphase I)
    • Failure of separation of sister chromatids (in anaphase II) **[Max 2]**

3. (a) I – aorta, II – pulmonary artery, III – pulmonary vein, IV – vena cava **[2]**
  (b) • Arteries have thicker wall/ veins have a thinner wall
    • Veins contain valves whereas arteries do not
    • Arteries have a narrower lumen than veins
    • Arteries have more muscle/elastic fibres than veins **[Max 3]**
  (c) Coronary artery **[1]**
  (d) • Clotting factors are released
    • Platelets form a plug (at the rupture site)
    • Inactive prothrombin is converted to the enzyme thrombin
    • Thrombin catalyses/converts fibrinogen to fibrin
    • Fibrin is insoluble and forms a mesh of fibres around the plug / fibrin traps blood cells forming a clot **[Max 3]**

4. (a) • Hormone leptin
    • Produced in adipose tissue
    • Acts on hypothalamus
    • Stomach has stretch receptors that send nerve impulses to the brain
    • Makes person feel full/satiated/decreases appetite
    • In overweight people the brain may be less sensitive/more tolerant to leptin **[Max 3]**
  (b) Removal of the waste products of metabolism / metabolic waste / metabolic activity **[1]**
  (c)

	Fish	mammals
Nitrogenous waste	Ammonia / NH3	Urea /
Toxicity	Higher	Lower
Water needed to dilute	More dilute	More concentrated
Energy required to make	Lower	Higher

[Max 2]

5. (a) Stem **[1]**
  (b) X – phloem – transports sugar/sucrose/amino acids/organic compounds
    Y – xylem – transports water (and minerals) **[2]**
  (c) • Phloem has smaller diameter tubes / sieve plates/more cytoplasm in cell/companion cells
    • Xylem has a thicker wall
    • Xylem has a wider lumen
    • Xylem as it is lignified
    • Xylem cells are dead whereas phloem are living **[Max 1]**
  (d) Hydrogen **[1]**
  (e) • Water molecules are polar/dipoles (so form hydrogen bonds)
    • Cohesion is when water molecules form (hydrogen) bonds with other water molecules
    • Adhesion is when water molecules form (hydrogen) bonds with other surfaces
    • Cohesion transmits the tension/ transpiration pull from the leaves to the stem to the roots/ maintains a continuous column of water
    • Adhesion is when water is attracted to xylem stopping column breaking / replaces water that has evaporated by adhesion with (cellulose) cell walls **[Max 3]**

Section B

6. (a)

	Globular	Fibrous
Shape	Compact/blob shape/ spherical	Long and thin/ parallel threads/ strands/sheets
Amino acids	More polar/soluble	Non-polar/insoluble
Function [any 2]	• Metabolic reactions (e.g. enzymes rubisco/ lysozyme/ amylase/catalase to catalyse reactions) • Immune response (e.g. immunoglobulins/ antibodies) • Transport oxygen (e.g. hemoglobin) • Hormone (e.g. epinephrine)	• Structural roles (e.g. collagen in connective tissue/skin, keratin in hair/ nails, fibrin for blood clots) • Movement (e.g. actin and myosin)
Sensitivity to pH and temperature	More sensitive/likely to denature	Less likely to denature
Solubility in water	Soluble	Insoluble

[Max 3]

  (b) • The final product of the pathway acts as a non-competitive inhibitor
    • The inhibitor binds to the allosteric site...
    • ...of the first enzyme in the pathway
    • The inhibitor changes the shape of the active site
    • The active site is no longer complementary to the substrate
    • The enzyme no longer binds to the substrate so no products are made
    • The metabolic pathway is turned off until the level of inhibitor decreases / inhibition is temporary
    • Decreasing the concentration of the inhibitor results in the inhibitor leaving the enzyme (allosteric site) and the pathway turns back on
    • End product inhibition prevents excessive production of a product...
    • ...by negative feedback
    • (E.g. isoleucine (the end product) binds to the threonine deaminase (enzyme that converts the substrate threonine to product) **[Max 4]**

  (c) • Gene regulation is the control of which genes on the DNA are expressed
    • In multicellular organisms, differentiation causes different genes to be expressed in different tissues
    • Most genes in DNA are not transcribed / are silent
    • Acetylation of histone proteins increases gene expression
    • Methylation of DNA/cytosine bases reduces/silences gene expression

- Supercoiled chromatin/heterochromatin has methylated DNA and deacetylated histones / relaxed chromatin/euchromatin has acetylated histones and demethylated DNA
- Open chromatin/euchromatin allows the binding of <u>transcription factors</u> increasing gene expression / supercoiled chromatin/heterochromatin prevents the binding of <u>transcription factors</u> decreasing gene expression
- Methylation/phosphorylation of <u>histone proteins</u> can alter gene expression
- Epigenetics – methylation patterns are inherited / increase during lifetime
- Environmental factors/drugs/alcohol/tobacco/cocaine can alter methylation of DNA
- Cell's hormones can alter gene expression by binding to (protein) receptors on cells causing the release of transcription factors
- Transcription factors can be activators or repressors / promote gene expression by increasing the binding of RNA polymerase to promoter / repressors decrease gene expression by preventing the binding of RNA polymerase to promoter
- Transcription involves the DNA acting as a template to produce mRNA
- Enhancers are sequences on the DNA which increase transcription of a gene when a protein binds to them / silencers are sequences on the DNA which decrease transcription of a gene when a protein binds to them
- mRNA can be altered after transcription/post-transcriptional modification / (e.g. removal of introns/alternative splicing/capping/addition of poly-A-tail)
- Micro RNAs can destroy mRNA / block mRNA so translation is inhibited
- Post-translational modification/phosphorylation can activate or deactivate proteins **[Max 8]**

7. (a) • A larger surface area of the lung is type I than type II
   - Type I cells are thinner squamous
   - Type I cells are adapted for a short diffusion distance/gas exchange
   - Type II cells have many mitochondria and Golgi body
   - Only type II cells secrete surfactant/lipids and proteins to decrease surface tension / stop alveoli sticking together
   - Type I cells are thicker/cuboidal as they are adapted to store phospholipids/surfactant
   - Type II cells are twice as common/more common than type I **[Max 3]**

(b) • Ventilation consists of inhalation and exhalation
   - Inhalation
     ○ External intercostal muscles contract and move up and out
     ○ The diaphragm contracts and moves down
     ○ The volume in the thorax increases...
     ○ ...reducing the pressure
     ○ This means the pressure is below atmospheric pressure
     ○ Air flows from the atmosphere into the lungs to equalize the pressure / from a high to low pressure
   - Exhalation
     ○ Exhalation involves relaxing of external intercostal muscles or contraction of internal intercostal muscles
     ○ The diaphragm relaxes and moves up...
     ○ ...increasing pressure in thorax
     ○ Air flows from the lungs to the atmosphere to equalize the pressure / from a high to low pressure
     ○ Forced exhalation involves contraction of internal intercostal muscles **[Max 5]**

(c) • Evolution is the change in heritable characteristics of a species over time
   - Mutations occur in the DNA of bacteria
   - Leading to variation in the resistance to antibiotics
   - The introduction of antibiotics alters the environment / causes a selection pressure
   - Bacteria not resistant to antibiotics are killed / the bacteria with the antibiotic-resistant mutation survive / survival of the fittest antibiotic-resistant bacteria
   - The fittest reproduce by binary fission
   - The population now all contain the antibiotic resistance **[Max 7]**

8. (a) Diagram could include:
   - Z lines
   - Actin filaments shown as thin lines attached to Z line
   - Myosin filaments with heads shown as thick lines in centre of sarcomere

- Light/I band and dark/A band identified

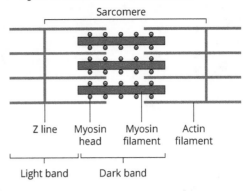

Z line   Myosin head   Myosin filament   Actin filament

Light band   Dark band **[Max 4]**

(b) • Skeletal muscle is made up of myofibrils
   - Myofibrils are made up of sarcomeres
   - <u>Proteins</u> are actin and myosin
   - Calcium ions released from sarcoplasmic reticulum...
   - ...bind to troponin which changes shape moving tropomyosin...
   - ...exposing myosin head binding sites on the actin filament
   - Myosin heads bind to actin filament forming cross-bridges
   - ATP binds to myosin heads and breaks cross-bridges
   - ATP is hydrolysed into ADP + P_i
   - Myosin head becomes cocked / changes configuration
   - Myosin head binds (myosin head binding site) to actin further from the centre of the sarcomere
   - P_i is released so the myosin head returns to its original shape/configuration
   - Actin filaments slide inwards towards centre of sarcomere / actin filaments slide past myosin filaments
   - Power stroke shortens sarcomere **[Max 6]**

(c) • Neonicotinoids are structurally similar to nicotine
   - Affect cholinergic synapses / synapses with acetylcholine as the neurotransmitter
   - Neonicotinoids bind irreversibly to acetylcholine receptors on postsynaptic membrane
   - Acetylcholinesterase cannot break down neonicotinoids
   - Causing continuous stimulation/overstimulation of the postsynaptic neuron
   - Causing convulsions/paralysis and death of insects
   - Neonicotinoids affect insects more than other animals as their acetylcholine receptors have a different structure / bind more strongly with neonicotinoids
   - Neonicotinoids kill non-target insects as well as pests
   - (E.g. honey bee populations decline / fewer insects to pollinate plants / aquatic insects die)
   - At low doses they reduce feeding / reproduction / immune system of insects/fish
   - Bird populations fall as fewer insects to feed on
   - Some countries have banned the use of neonicotinoids **[Max 5]**

Set C: Paper 3

Section A

1. (a) 0.64 **[Accept between 0.63–0.65]**
   (5.6/8.8) **[1]**
   (b) Any organic solvent e.g. ethanol/acetone/hexane/pentane **[1]**
   (c) Fucoxanthin / Chlorophyll c **[1]**
   (d) • Both absorb well in the red and blue area of the spectrum
   - Both do not absorb well in the green area of the spectrum
   - At 400 nm/violet/indigo chlorophyll a absorbs more light than chlorophyll b
   - At 450 nm/indigo chlorophyll b absorbs more light than chlorophyll a
   - At 625 nm/orange chlorophyll b absorbs more light than chlorophyll a
   - At 675 nm/red chlorophyll a absorbs more light than chlorophyll b
   - Chlorophyll b has the highest peak absorbance in the blue end of the spectrum
   - Chlorophyll a has the highest peak absorbance in the red end of the spectrum **[Max 2]**
   (e) • Fucoxanthin absorbs light better than chlorophyll a and b in the blue green / 475–550 part of the spectrum
   - Fucoxanthin absorbs light better than chlorophyll a and b in the violet and indigo / 400–450 part of the spectrum

- Increased photosynthesis in the blue end of the spectrum
- Increased photosynthesis when underwater as can use the green light **[Max 1]**

2. (a) Potometer **[1]**
(b)
- Size of plant/same cutting
- Surface area of leaves
- Light level
- Temperature
- Humidity **[Max 1]**
(c) To allow the plant to acclimatize **[1]**
(d) distance bubble moved per minute = (34 − 0)/10 = 3.4 mm/min **[1]**
(e)
- Method assumes the rate of water transpired is equal to the rate of water taken up
- Some water is used in photosynthesis/some is made in respiration so data may be inaccurate **[Max 1]**
(f)
- Measured diameter = 35,000 µm
- Scale bar image size = 14 ÷ 15 mm = 14,000 ÷ 15,000 µm
- Actual size of X = 35,000/14,000 × 5
- Actual size = 5 µm
- 12.5 µm **[Accept 11.75–12.75 µm] [1]**

3. (a) Amino acids / glycine / name of any amino acid **[1]**
(b)
- Organic molecules/amino acids could be produced by chance
- Without having been made by other living molecules **[1]**
(c)
- Small scale so may not have been same result on larger scale
- We don't know for certain the exact conditions in early Earth **[Max 1]**

Section B

Option A: Neurobiology and behaviour

1. (a) Broca's area in the left cerebral hemisphere **[1]**
(b) Functional fMRI / functional Magnetic Resonance Imaging **[1]**
(c)
- Damage to Broca's area normally results in the inability to convert thoughts to speech
- This person is however able to convert thoughts to speech
- The right-hand part of the brain lit up has taken over the job of Broca's area
- Neural plasticity has occurred **[Max 2]**

2. (a)
- I – rod cell
- II – cone cell
- III – bipolar neuron/cell
- IV – ganglion cell/neuron

  [1 mark for 2 correct answers: 2 max]
(b)
- In the dark rod and cone cells become depolarized / −50 mV
- Depolarization releases an inhibitory neurotransmitter/ glutamate into the synapse with the bipolar cell
- The inhibitory neurotransmitter binds to the postsynaptic membrane/bipolar cell membrane
- The bipolar cell becomes hyperpolarized / −80 mV
- Hyperpolarized cell is less likely to generate an action potential to its ganglion cell
- In light the rod and cone cells are hyperpolarized so transmit less inhibitory neurotransmitter/glutamate
- In light the bipolar cell is not inhibited so generates an action potential
- The action potential passes to the ganglion cell
- Ganglion cells increase the frequency of nerve impulses/ action potentials along axons
- Axons all meet at the blind spot
- The impulses are sent along the optic nerve through the optic chiasma to the visual cortex of the brain **[Max 4]**

3. (a)
- Aging decreases the synthesis of the protein GAD67
- Less GABA is synthesized
- Gene expression of GAD67 / GABA is reduced **[1]**
(b) Inhibitory (postsynaptic potential) as neuron is hyperpolarized **[1]**
(c)
- Aging increases the resting potential / the resting potential is less negative with age / −70 mV to −63 mV
- Aging makes neurons more likely to reach threshold potential/depolarize **[2]**
(d)
- Inhibitory synapses use GABA as their neurotransmitter
- Less GABA is made/synthesized in the presynaptic neuron / less GABA is released by exocytosis from presynaptic neuron
- GABA diffuses across the synapse and binds to receptors on the postsynaptic membrane
- Fewer chloride channels open / causing the postsynaptic neuron to be less hyperpolarized

- The postsynaptic neuron is less negative / more likely to depolarize
- Acoustic information cannot be processed so easily **[3]**

4. (a) Foraging **[1]**
(b) The longer the shell the thicker the shell / positive correlation / as shell length increases, shell thickness increases **[1]**
(c)
- As shell size/thickness increased the proportion of mussels opened compared with the mussels present decreases (or numerical example)
- The difference is significant above 60 ×10^{-2} / 0.6 mm as the error bars do not overlap between number opened and number present
- The difference is not significant at approximately 50 ×10^{-2} / 0.5 mm as the error bars do not overlap between number opened and number present **[Max 2]**
(d)
- 52 mm long **[Accept any value in range 50–54]**
- As they are the most profitable in terms of energy yield obtained from food for the time spent foraging
- Larger shells supply more food / small shells don't supply enough food/energy
- But larger shells are difficult to crack open / take more energy to open
- The shells above 60 are too difficult to crack open **[Max 3]**
(e)
- Sound waves are funnelled through the pinna
- Sound waves cause the ear drum / tympanic membrane to vibrate
- Bones of the middle ear / ossicles/hammer anvil and stirrup / malleus, incus and stapes amplify the movement
- The stapes/stirrup transmit sound waves the oval window vibrates
- Fluid inside the cochlea vibrates
- The round window vibrates in the opposite direction to the eardrum allowing the fluid in the cochlea to vibrate
- Hairs in the cochlea are mechanoreceptors and vibrate
- Different hairs vibrate at specific frequencies
- The hairs release a neurotransmitter
- The neurotransmitter generates an action potential
- The nerve impulse travels to the brain along the auditory nerve **[Max 6]**

Option B: Biotechnology and bioinformatics

1. (a) Leptin reduces appetite as the leptin-deficient mouse model is obese **[1]**
(b) **Advantages**
- No leptin gene so the function of the leptin gene can be deduced by comparing with a control mouse
- Could lead to improvements in medicine for people
- It is unethical to test on people before it has been shown to be effective on animals
- The mouse is a frequently used model organism so gene function in mice can be a good predictor in humans **[Max 3]**
Disadvantages
- Mice may not respond in the same way as humans
- Ethical issues associated with animal welfare **[2]**
(c)
- Knockout mice are genetically modified
- To be missing a gene of interest
- The function of the gene can be deduced from the change to the phenotype
- Advantage is the mice can be made with the researcher's specific gene missing
- Rather than relying on mice that have been selectively bred for research
- E.g. mice with leptin gene knocked out are obese so leptin has a role in reducing obesity. **[Max 3]**

2. (a)
- Gene therapy involves inserting healthy genes to replace defective genes
- Healthy ADA gene is extracted from a healthy person and inserted into a disabled retrovirus
- The retrovirus/virus is used as a vector containing the human ADA gene
- Haemopoietic/stem cells are collected from the patient's bone marrow / umbilical cord
- Retrovirus/virus inserts the ADA/gene into chromosomes in the stem cells
- Transgenic stem cells are cultured and injected into the patients' blood
- Stem cells move to bone marrow and make T cells **[Max 4]**

(b) • There is an increased risk of leukaemia cancer as stem cells divide repeatedly
  • Treatment only lasts a few years then needs to be repeated
  • Host may develop immunity to the virus so vector may not be useful a second time
  • Trials of gene therapy on humans have resulted in death of participants
  • Somatic gene therapy cannot pass on genetically modified/ transgenic cells to offspring
  • Germline gene therapy risks altering the human genome in generations to come **[Max 3]**

(c) • mRNA of the gene is extracted and used as a template to make cDNA / a DNA strand
  • Using the enzyme reverse transcriptase
  • ESTs are used in data mining to identify similar genes with a known function
  • ESTs can determine the position of a gene on a chromosome **[Max 2]**

3. (a) • The genus and species are given helping with taxonomy/ evolutionary relationships
  • International cooperation needs a common language / different cultures may have different common names **[Max 1]**

(b) Phylogram, as the length of each branch is proportional to the difference between the organisms **[1]**

(c) *Felis catus* and *Panthera pardus* as the length of the branch is shorter **[1]**

4. (a) 46 M/46,000,000
  [Accept anywhere between 46 M and 47 M or 46 million to 47 million] [1]

(b)

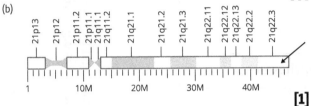

[1]

(c) • p – short arm
  • q – long arm **[1 for both correct]**

(d) • Databases allow for sharing of information across the scientific community / international cooperation / collaboration
  • Many are free / easy to access
  • Allows faster data collection due to improvements in comparing and searching genetic sequences
  • The body of data in databases is increasing exponentially
  • BLAST searches identify similar genetic sequences in different organisms e.g. nucleotide sequences or protein sequences / Phylogenetics uses multiple sequence alignment
  • Databases can be searched to compare new genetic sequences to other known sequences e.g. SARS-COV-2
  • Gene function can be investigated in model organisms
  • Sequences can be compared between different species
  • EST is an expressed sequence tag that can be used to identify possible genes
  • Databases allow for quicker research and development of vaccines
  • However, some private companies do not make DNA sequences public so they can develop a product and make a profit
  • Example of 2 bioinformatics databases and their use from:
    o Uses shared language e.g. FASTA is the format used to search nucleotide or protein databases with a query sequence
    o Genetic databases e.g. GenBank publish sequences of all publicly available DNA sequences
    o Nucleotide sequencing databases e.g. EMBL/ European Molecular Biology Laboratory /
    o The National Centre for Biotechnology NCBI provides biomedical and genomic information. /
    o RCSB – protein data bank archive provides information about the 3D shapes of proteins, nucleic acids, and complex assemblies
    o SwissProt – protein sequencing database resource of protein sequence and functional information
    o OMIM online catalogue of human genes and genetic disorders
    o GenAge Human Genes Chromosome locations

o Ensembl shows many animals including human chromosomes and location of genes
    o 3D structural databases e.g. Protein data bank
    o BLAST / Basic Local Alignment Search Tool can compare sequences between species
    o Microarray databases e.g. Array express
    o Pathway databases e.g. Kyoto Encyclopaedia of Genes and Genomes **[Max 6]**

Option C: Ecology and conservation

1. (a) Alien species **[1]**
(b) • Interspecific competition occurs in all four countries
  • The population has drastically decreased more in Wales and England as the grey squirrel has outcompeted the red squirrel
  • Ireland and Scotland still have large areas without competition from the grey squirrel **[Max 2]**

(c) • Less habitat is lost in Scotland and Ireland / more conifer trees / protect habitat
  • Conservation efforts to protect the red squirrel from the grey squirrel
  • Culling of grey squirrels to prevent the spread of poxvirus **[Max 1]**

(d) • The red squirrel reduces the number of visits after pine marten scent but the grey squirrel increases visits
  • The red squirrel spends significantly more time being vigilant than the grey squirrel after the pine marten scent has been added **[Max 1]**

(e) Biological control **[1]**
(f) Arguments for:
  • The red squirrel is a native species so has evolved to react to pine marten scent/avoid pine martens/be more vigilant so is unlikely to be eaten
  • The grey squirrel is not a native species so has no response to the pine marten so is more likely to be eaten
  • The study is a large sample size so is representative of squirrels and pine marten interaction
  • If more grey squirrels are eaten that reduces interspecific competition so the red squirrel numbers could recover
  Arguments against:
  • The grey squirrels could develop a response to pine marten scent over time
  • Even if grey squirrels disappear there are few places with enough conifers to support the population of red squirrels
  • The study was carried out in Northern Ireland so may not be representative of squirrels in England **[Max 2]**

2. (a) Autotrophs **[1]**

(b)
```
                    shrimp
phytoplankton → zooplankton → planktivorous fish
algae
```

[2 marks for correct food web]
[1 mark for: phytoplankton → zooplankton → planktivorous fish → piscivorous fish]
[1 mark for: algae → zooplankton → shrimp]

(c) A factor that restricts the growth of a population when in short supply **[1]**

(d) • Top-down: increased zooplankton
  • Bottom-up: availability of nutrients/nitrogen/phosphate **[Max 1]**

(e) • NOx from burning fossil fuels dissolves in water
  • Phosphates/nitrates from sewage / fertilizer
  • Phosphates from detergents
  • Monoculture/deforestation increases runoff of nutrients from land **[Max 1]**

(f) • Increased nutrients/nitrogen/phosphate increase growth/ population of primary producers/algae/phytoplankton/algal bloom
  • Increases population of zooplankton/primary consumers/ consumers
  • Algal bloom/algae block sunlight
  • Algae/plants die and decomposition increases
  • Increased BOD/biological oxygen demand/respiration causes oxygen levels in the water to decrease / lake becomes anoxic
  • Biodiversity/fish/aquatic organisms decrease **[Max 4]**

3. (a) Primary and secondary consumers **[1]**
(b) • Top-down: increase in *Larus argentatus*
  • Bottom-up: decreased population of *Semibalanus balanoides* **[1 mark for both]**

(c) • Capture–mark–release–recapture / Lincoln index

167

- Mark out a set area
- Capture every individual in the area and mark them with a spot
- The mark must not harm the animal e.g. non-toxic nail varnish
- Return to the same area after a suitable amount of time / a day
- Recapture all organisms in the area

- Calculate the population size by the equation
- Population size = $\dfrac{n1 \times n2}{n3}$

(n1 = original number captured/marked, n2 = total number recaught, n3 = number of recaptured that were marked) **[Max 4]**

4. **[1 mark for every 2 boxes linked by a line with the line labelled correctly]**

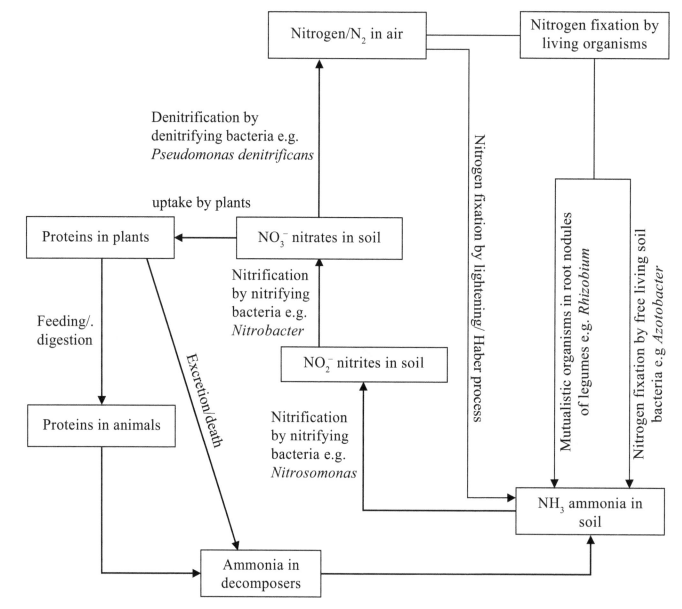

Option D: Human physiology

1. (a) • In each population an increase in cholesterol increases the risk of CHD mortality
  - Positive correlation   **[1]**
  (b) Northern Europe   **[1]**
  (c) • Southern Europe mediterranean has a lower CHD mortality risk than Southern Europe inland
  - Both have serum cholesterol levels between 3.65 and 6.50 / similar range of cholesterol
  - Both show an increase in cholesterol increases death rate by CHD
  - (below 6.5 mmol) The risk of dying from CHD is approximately double in inland compared with mediterranean   **[Max 2]**
  (d) • Mediterranean diet has a higher proportion of fibre
  - Mediterranean diet has a higher proportion of unsaturated fat
  - Mediterranean diet has a lower LDL cholesterol/higher HDL cholesterol
  - Mediterranean diet has a higher proportion of unsaturated fat/ lower proportion of saturated/trans fats
  - Oils such as olive oil and fish oils are high in healthy omega 3 fatty acids/lower in omega 6
  - Differences in genetics between the populations
  - Differences in obesity levels
  - Differences in exercise between the populations
  - Differences in smoking between the populations   **[Max 1]**

  (e) • There is a positive correlation between cholesterol and CHD
  - However, correlation does not imply causation
  - Many other factors affect CHD / example of factor
  - Only high (LDL) cholesterol results in fatty deposits/atheroma in arteries but high HDL cholesterol reduces CHD   **[Max 2]**

2. (a) • X = Aortic pressure
  - Y = Ventricular pressure   **[2]**
  (b) 200 ms   **[1]**
  (c) • To ensure the atria have fully contracted before the ventricles start to contract
  - To ensure the atria have fully contracted before the atrioventricular valves shut
  - To ensure the blood in the atria has passed into the ventricles   **[Max 1]**
  (d) • Valves open and close in response to changes in blood pressure/heart contraction/pumping
  - Valve prevents backflow/maintains direction of blood flow
  - Valves allow heart chambers to fill/to empty   **[2]**

3. (a) 104 mmHg = 97/98% **[1]**

(b)
- There is a concentration gradient / higher partial pressure inside the alveoli than inside the capillary/inside the alveoli is 104 mmHg and the capillary is 40mmHg
- Oxygen moves out of the alveoli into the capillary
- Down the concentration gradient/ from higher to lower concentration
- By <u>diffusion</u>
- Through the pneumocytes/basal lamina/capillary endothelium **[Max 2]**

[Max 3]

(c) <u>Hemoglobin</u>
- The curve is sigmoid as the binding of each hemoglobin subunit with each O_2 changes the conformation/shape/charge making it easier for hemoglobin to bind to the next oxygen / co-operative binding of each hemoglobin subunit to each oxygen occurs
- Sigmoid shape as there are 4 polypeptides and can carry up to 4 molecules of O_2
- Hemoglobin has a higher affinity for oxygen binding at high partial pressures so binds with more oxygen in the lungs
- Hemoglobin has a lower affinity for oxygen binding at low partial pressures so releases oxygen in the tissues

Myoglobin **[Max 3]**
- Has a higher affinity for oxygen than hemoglobin
- Will only release oxygen when the hemoglobin supply is exhausted delaying the onset of anaerobic respiration/ lactic acid formation
- In myoglobin there is no co-operative bonding giving a linear shape
- In myoglobin there is only 1 subunit so myoglobin will not release oxygen, the partial pressure in the muscle tissue is very low **[Max total: 4]**

4. (a) Endocrine **[1]**

(b)

Comparison	Peptide hormone	Steroid hormone
Solubility	Hydrophilic / water soluble	Hydrophobic / fat soluble
Can diffuse through lipid bilayer/plasma membrane of target cell	No	Yes
Bind to receptor	Receptor protein on cell membrane	Steroid receptor in cytoplasm
Affects transpiration of specific mRNA/ gene expression	Indirectly via relay molecules/signal transduction/ stimulates relay messengers to take message to nucleus	Complex directly binds to chromatin

Comparison	Peptide hormone	Steroid hormone
Secondary messengers activated/signal transduction	Yes, e.g. cyclic AMP/ protein kinases/ calcium ions/nitric oxide	No
Amplification	Yes, as can cause a cascade of intracellular secondary messengers inside the cell	No, as it directly acts on chromatin

[Max 3]

5. The lungs, kidneys and amino acids/buffers are all involved in regulating blood pH
Increased exercise increases respiration so carbon dioxide/CO_2 levels increasing the acidity of the blood

Bicarbonate buffer
- Carbon dioxide/CO_2 dissolves in blood plasma forming carbonic acid/H_2CO_3
- Carbonic acid/H_2CO_3 acts as a bicarbonate buffer / If the blood is too acidic H_2CO_3 dissociates into acidic ions/H^+ and alkaline ions/ hydrogen carbonate ions/HCO_3^-
- <u>Chemoreceptors</u> in aortic bodies/carotid bodies/medulla/ respiratory centre monitor the blood pH/detect the higher concentration of H^+ ions
- Chemoreceptors/receptors send nerve impulses to respiratory centre to the diaphragm / intercostal muscles increase ventilation rate/depth to lower the concentration of CO_2 in the blood
- Less CO_2 causes the bicarbonate buffer to reform H_2CO_3 from plasma H^+ and HCO_3^-
- Less H^+ ions are in the plasma so the pH rises/becomes less acidic

Kidney [Max 1]
- Excess H^+ ions from the blood are excreted in the urine
- Less hydrogen carbonate ions/HCO_3^- ions from the blood are excreted in the urine
- (Raising) the blood pH as there are relatively more basic ions/ HCO_3^-

Amino acids [Max 1]
- Amino acids are zwitterions / have a positive/NH_3^+ end and a negative/COO^- end
- In excess acid the extra protons/H^+ are mopped up by the COO^- end forming COOH (raising the blood pH)
- The excess protons/hydrogen ions/H^+ are slowly excreted by the urine **[Max 6]**

NOTES

CPSIA information can be obtained
at www.ICGtesting.com
Printed in the USA
BVHW012330171122
652262BV00010B/319